Collins

AQA GCSE 9-1

Combined Science

Trilogy

Combined Science

Foundation

AQA GCSE 9-1

Workbook

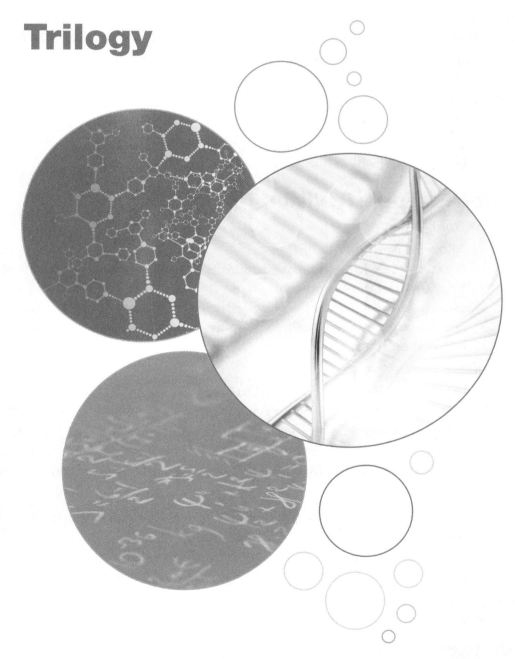

Ian Honeysett, Gemma Young
and Nathan Goodman

Revision Tips

Rethink Revision

Have you ever taken part in a quiz and thought *'I know this!'* but, despite frantically racking your brain, you just couldn't come up with the answer?

It's very frustrating when this happens but, in a fun situation, it doesn't really matter. However, in your GCSE exams, it will be essential that you can recall the relevant information quickly when you need to.

Most students think that revision is about making sure you **know** stuff. Of course, this is important, but it is also about becoming confident that you can **retain** that *stuff* over time and **recall** it quickly when needed.

Revision That Really Works

Experts have discovered that there are two techniques that help with all of these things and consistently produce better results in exams compared to other revision techniques.

Applying these techniques to your GCSE revision will ensure you get better results in your exams and will have all the relevant knowledge at your fingertips when you start studying for further qualifications, like AS and A Levels, or begin work.

It really isn't rocket science either – you simply need to:

- **test yourself** on each topic as many times as possible
- **leave a gap** between the test sessions.

Three Essential Revision Tips

1. **Use Your Time Wisely**

   - Allow yourself plenty of time.
   - Try to start revising at least six months before your exams – it's more effective and less stressful.
   - Your revision time is precious so use it wisely – using the techniques described on this page will ensure you revise effectively and efficiently and get the best results.
   - Don't waste time re-reading the same information over and over again – it's time-consuming and not effective!

2. **Make a Plan**

   - Identify all the topics you need to revise.
   - Plan at least five sessions for each topic.
   - One hour should be ample time to test yourself on the key ideas for a topic.
   - Spread out the practice sessions for each topic – the optimum time to leave between each session is about one month but, if this isn't possible, just make the gaps as big as realistically possible.

3. **Test Yourself**

   - Methods for testing yourself include: quizzes, practice questions, flashcards, past papers, explaining a topic to someone else, etc.
   - Don't worry if you get an answer wrong – provided you check what the correct answer is, you are more likely to get the same or similar questions right in future!

Visit our website for more information about the benefits of these techniques and for further guidance on how to plan ahead and make them work for you.

www.collins.co.uk/collinsGCSErevision

Contents

Contents

Contents

Cell Structure

1 **Figure 1** shows an animal cell.

Figure 1

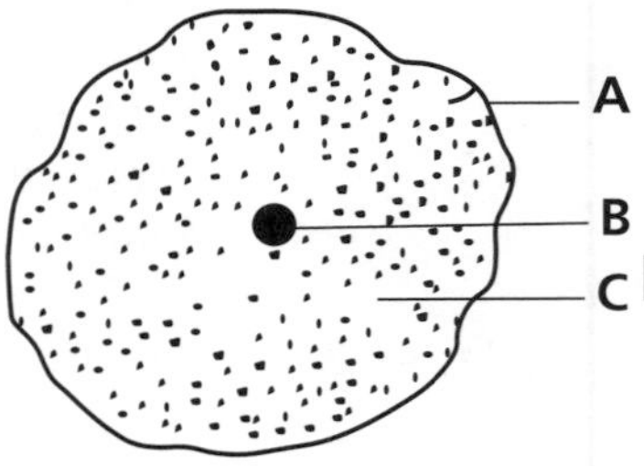

Which part of an animal cell do the following descriptions refer to? Write the letter **A**, **B** or **C** from the diagram next to each statement.

a) The part of the cell that contains DNA. ....

b) The part of the cell that controls the movement of

substances in and out. ....

[1]

c) The part of the cell where most of chemical reactions take place. .... **[1]**

2 **Figure 2** shows two different types of single-celled organism.

Figure 2

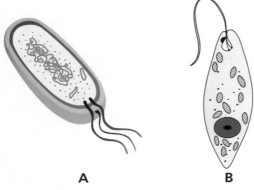

A   B

a) Which of the cells is eukaryotic? Explain how you can tell.

....

.... **[2]**

b) The two diagrams are not drawn to the same scale.

How can you tell this?

.... **[1]**

c) The two cells move in a similar way.

Describe how they can move.

....

.... **[2]**

Total Marks .... / 8

Investigating Cells

1 Rajesh uses his light microscope to view and draw a cheek cell.
He puts some cheek cells on to a slide and adds a few drops of a blue solution.

Figure 1 shows what Rajesh draws.

Figure 1

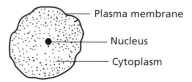

Plasma membrane

Nucleus

Cytoplasm

a) Why does Rajesh add a few drops of blue solution to the cells?

_____

_____ **[2]**

b) A cheek cell is actually 0.03mm in diameter.

   i) What is the diameter in micrometres?

   Answer: _____ **[1]**

   ii) Calculate the magnification of Rajesh's drawing.

   $$\text{magnification} = \frac{\text{size of image}}{\text{size of real object}}$$

   Answer: _____ **[2]**

2 Both the light microscope and the electron microscope are used to look at cells.

Put a tick (✔) or a cross (✗) in each box in the table to show the features of the two different types of microscope.

	Light Microscope	Electron Microscope
Can be used to see both animal and plant cells.		
Works by passing light through the specimen.		
Can be used to see the detailed structure of mitochondria.		

[3]

Total Marks _____ / 8

Cell Division

1   A student investigates the growth of onions.
    He puts an onion bulb in a jar of water.
    The bulb starts to grow roots.

    a) Cells in the root tip are dividing.

        Which part of the cell cycle involves cells dividing?

        .... [1]

    b) The student draws a diagram of a cell that is dividing from a fruit fly.
        Their diagram is shown in **Figure 1**.

        Figure 1

        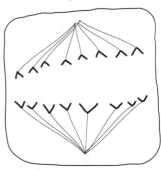

        Describe what is happening inside the cell.

        ....

        ....

        .... [2]

2   Read the following quote and answer the questions below.

    "People are always against new ideas such as using stem cells, but within a few years they
    will be used all the time to cure diseases."

    a) How can stem cells be used to cure diseases?

        ....

        .... [2]

    b) Give **one** reason why people may be against using stem cells to cure diseases.

        .... [1]

Total Marks .... / 6

Transport In and Out of Cells

1 The three blocks in **Figure 1** were cut from a block of agar jelly that had been dyed green.

Figure 1

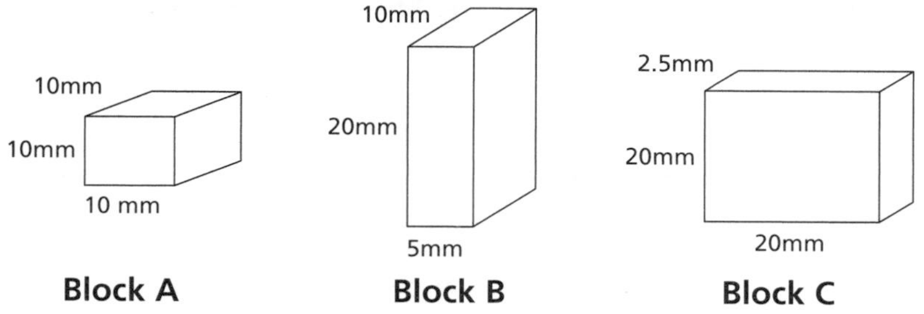

Block A — 10mm, 10mm, 10 mm

Block B — 10mm, 20mm, 5mm

Block C — 2.5mm, 20mm, 20mm

Block A   **Block B**   **Block C**

a) Calculate the surface area of block **A**. Put the results in **Table 1**.

Table 1

Block	Surface Area (mm²)	Volume (mm³)
A		1000
B	700	1000
C	1000	1000

[1]

b) The green dye turns yellow when acid moves through into the agar jelly.

What process would cause the acid to move through the agar jelly?

Answer: _____ [1]

c) The three blocks are submerged in an acid solution.

Which block would be the first to change colour completely? Explain your answer.

_____

_____

_____

[3]

2 Give **two** differences between:

a) Osmosis and diffusion.

_____

_____ [2]

b) Diffusion and active transport.

_____

_____ [2]

Total Marks _____ / 9

Levels of Organisation

1 **Systems, cells, organs** and **tissues** are all levels of organisation in living organisms.

a) Write down these four levels in order of complexity, with the least complex first.

_____ **[1]**

b) Give the level of organisation of each of these structures.
The first one has been done for you.

A lymphocyte	*cell*
The heart	
A neurone	
A leaf	
Epithelia on the skin	

[4]

c) **Table 1** shows the number of mitochondria in different types of cell.

Which statement could best explain the data in the table?
Tick **one** box.

Table 1

Type of Cell	Number of Mitochondria
liver	1500
heart muscle	5000
skin	100

Heart muscle cells are specialised to make protein. ☐

Skin cells need large amounts of energy. ☐

Liver cells do not require much energy. ☐

Muscle cells are specialised for contraction. ☐   **[1]**

2 The following sentences explain how xylem and phloem cells are specialised for their jobs.

Put a ring around the word that best completes each sentence.

Xylem cells contain rings of **cellulose / cytoplasm / lignin** that make them strong.

Xylem cells do not have end walls, so **cytoplasm / sugars / water** can easily flow through them.

Phloem cells have **holes / nuclei / mitochondria** in their end walls to let dissolved substances through.   **[3]**

Total Marks _____ / 9

Digestion

1 The enzyme amylase breaks down starch.
The graph in **Figure 1** shows the effect of temperature on the enzyme amylase.

Figure 1

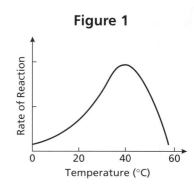

a) Describe the pattern shown in the graph.

....

.... **[2]**

b) Give the optimum temperature of this enzyme.

.... **[1]**

2 Lipids (fats) are digested in the body by the enzyme lipase.

a) Give one place in the body where lipase is made.

Answer: .... **[1]**

b) Some people want to be able to eat foods containing fat without gaining weight.
One way to do this is to replace fats in food with a substance called olestra.
Figure 2 shows an olestra molecule.

Figure 2

Fatty acid units

sugar units

Lipase digests lipids but cannot digest olestra.

Put a tick (✓) in the box next to the most likely reason for this.

Olestra molecules contain fatty acids but lipids do not.	
Lipid molecules dissolve in water.	
Olestra molecules are the wrong shape.	

[1]

Total Marks .... / 5

Blood and the Circulation

1 Match the numbers on **Figure 1** with the blood vessels listed below.
Write the appropriate numbers in the boxes provided.

Figure 1

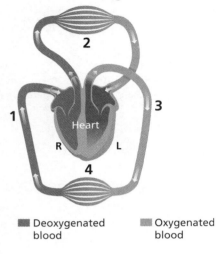

■ Deoxygenated blood   ■ Oxygenated blood

Aorta ☐

Vena cava ☐

Capillaries in the lungs ☐

Capillaries in the body ☐ **[3]**

2 Haemoglobin is found in red blood cells.

a) Explain how haemoglobin supplies the tissues of the body with oxygen.

....

....

.... **[3]**

b) Carbon dioxide is also transported in the blood.

i) By which part of the blood is most carbon dioxide transported?
Tick **one** box.

Red blood cells ☐   Plasma ☐

White blood cells ☐   Platelets ☐ **[1]**

ii) The blood takes up carbon dioxide in the tissues.

Where does the blood carry the carbon dioxide to?

.... **[1]**

Total Marks .... / 8

Non-Communicable Diseases

1 A build-up of fatty material in the wall of the coronary arteries of the heart is called atheroma or atherosclerosis.
It has been known for many years that atheromas can cause a heart attack.

a) i) What is a heart attack?

_____ **[1]**

ii) How can an atheroma cause a heart attack?

_____

_____ **[2]**

b) People used to think that the only cause of heart attacks was too much saturated fat, causing atheromas.

A scientist looked at groups of men who did different jobs involving different amounts of physical activity.
He recorded how many of these men suffered from heart attacks and how many had large atheromas. His results are shown on the graphs in **Figure 1**.

Figure 1

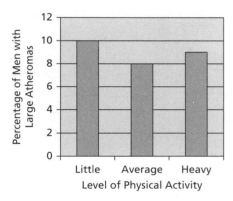 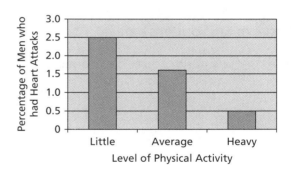

What conclusions can you make from these results?

_____

_____

_____ **[3]**

Total Marks _____ / 6

Transport in Plants

1   A student sets up an investigation into water transport in a plant as shown in **Figure 1**.
She measures the mass of the test tube and contents at the start and
again the next day.

She repeats the experiment with the plant under different conditions.

The student's results are shown in **Table 1**.

Figure 1

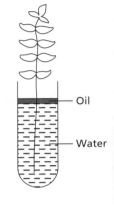

Oil

Water

Table 1

Conditions	Mass of Tube and Contents (g)	
	At Start	After 1 Day
normal	29.7	26.5
windy	30.4	25.5
humid	29.2	27.2

a)   Complete the following sentences about the student's results.

The plant loses mass because it loses _____ by a process called

_____ .

The greatest mass is lost from the plant in _____ conditions.   **[3]**

b)   Predict what the student's results would be under the following conditions compared
with the normal conditions.
You must explain your answers.

i)   Some of the smaller roots have been removed from the plant.

_____

_____   **[2]**

ii)   Some of the leaves are painted with nail varnish.

_____

_____   **[2]**

2   Give **two** differences between the movement of water and the movement of sugars in a plant.

_____

_____   **[2]**

Total Marks _____ / 9

Pathogens and Disease

1 a) Draw **one** line from each disease to the type of microorganism that causes it.

Disease	Type of Microorganism
rose black spot	protozoan
salmonella	fungus
measles	virus
malaria	bacterium

[3]

b) Complete the sentences using the words from the list.

host   *Anopheles*   *Plasmodium*   **protozoan**   **red**   **vectors**

Mosquitoes are _____ because they carry malaria.

The mosquito's blood carries the malaria pathogen, which is called _____ .

The mosquito is a parasite because it feeds on a living organism,

called the _____ .

[3]

2 Human Immunodeficiency Virus (HIV) is a pathogen that can cause AIDS.

a) Give **two** ways that HIV can be passed on from one person to another.

_____

_____

[2]

b) AIDS stands for Acquired Immunodeficiency Syndrome.

Why is it called 'immunodeficiency'?

_____

_____

[2]

3 A gardener discovers that one of his rose bushes has black spot disease.
A friend advises him to burn any infected leaves rather than put them in the compost heap.

Why does she suggest this?

_____

_____

[2]

Total Marks _____ / 12

Human Defences Against Disease

1 **Figure 1** shows some of the main ways that pathogens can enter the body.

Figure 1

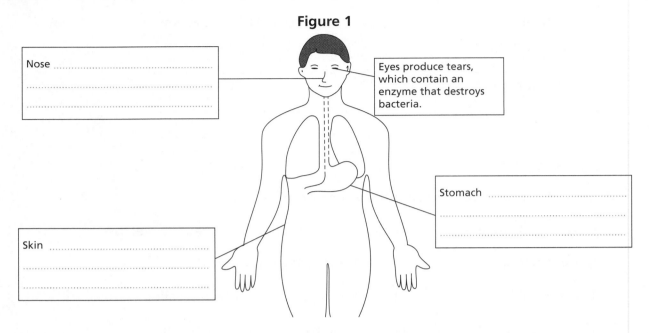

Nose ....
....
....

Eyes produce tears, which contain an enzyme that destroys bacteria.

Stomach ....
....
....

Skin ....
....
....

The diagram shows how the eyes work to stop pathogens entering the body.

Complete the diagram to show how the other **three** areas labelled prevent infection.   **[3]**

2 **Figure 2** shows how the body responds to a pathogen entering the bloodstream.

Figure 2

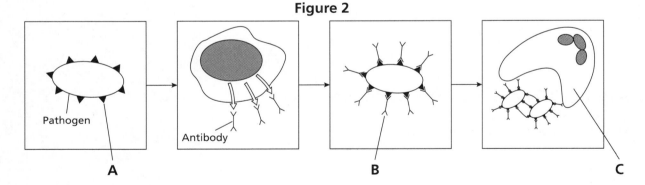

Pathogen   Antibody

A   B   C

Write down the names of structures **A**, **B** and **C** in the diagram.
Choose your answers from the following words.

antibody   antigen   phagocyte   platelet

A = ...

B = ...

C = ...   **[3]**

Total Marks .... / 6

Treating Diseases

1 This article is about the treatment of tuberculosis.

Antibiotic Could Beat TB

Scientists believe that they have found an antibiotic that could beat tuberculosis (TB).

TB is a disease of the respiratory system and is caused by a bacterium.

TB almost disappeared but is now killing more people. This is because the antibiotics that were used to treat TB are not effective now.

The new antibiotic is called linezolid and is being tested in trials.

---

a) People who have TB can be treated with antibiotics, but HIV cannot be treated in this way.

Give the reason why.

_____ [2]

b) Explain why TB is now on the increase.

_____

_____ [2]

c) Describe two precautions that can be taken when using antibiotics to help prevent problems like this occurring.

_____

_____ [2]

d) The article says that the new drug is being tested in trials.

Explain why drugs are tested in trials before they are given to large numbers of patients.

_____

_____

_____ [2]

Total Marks _____ / 8

Photosynthesis

1 Plants make glucose by photosynthesis.

a) Write down the word equation for this process.

_____ [2]

b) i) Glucose is transported from the leaves to other parts of a plant.

Suggest **two** parts of a plant that the glucose might be taken to.

_____

_____ [2]

ii) The glucose can also be built-up into different substances.
These substances can then be used in many different ways.

Name **two** of these substances and explain how they are used in a plant.

_____

_____

_____

_____ [4]

2 The passage is about early investigations into plant growth.

> In the seventeenth century, the biologist, Jan Baptiste van Helmont, grew a tree in a bucket of soil. He fed the tree with rainwater only.
>
> In five years, the tree had grown but the amount of soil in the bucket did not decrease. Baptiste concluded that extra material in the tree must have come from the rainwater.

Jan Baptiste van Helmont's conclusion was only partly correct.

Explain why.

_____

_____

_____ [2]

Total Marks _____ / 10

Respiration and Exercise

1 When Joanne runs a race her muscles work hard.
Her muscles use aerobic respiration to release energy from glucose.

a) Complete the word equation for aerobic respiration.

glucose + ... → .... + ... [3]

b) Joanne is training to run a
marathon.

When she runs, her muscles start
to make lactic acid and this passes
into her blood.

The graph in **Figure 1** shows the
lactic acid concentration in Joanne's
blood during her first training run
and during the first part of the race.

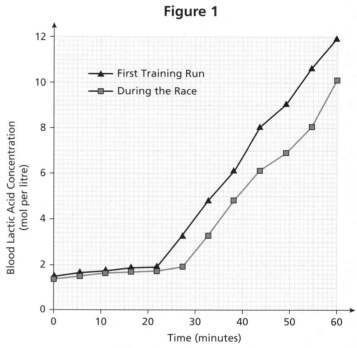

Figure 1

i) What is the concentration of lactic acid in Joanne's blood before she starts running
the race?

Answer: .... [1]

ii) Use the graph to explain why Joanne can run more efficiently when she has trained
for a race.

....

....

....

....
[3]

Total Marks .... / 7

Homeostasis and the Nervous System

1 Billy picks up a hot plate and quickly drops it without thinking about it.

a) How can you tell that this is a reflex action?

....

.... [2]

b) What type of cell detects the hot plate?

Answer: .... [1]

2 **Figure 1** shows part of the nervous system.

Figure 1

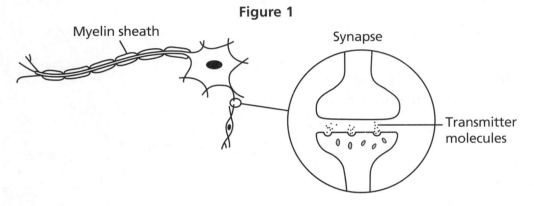

Myelin sheath

Synapse

Transmitter molecules

The enlarged section shows a synapse. Transmitter molecules carry signals across the synapse.

a) Name the type of cell shown in **Figure 1**.

.... [1]

b) Describe the function of the transmitter molecules at a synapse.

....

....

....

.... [3]

Total Marks .... / 7

Hormones and Homeostasis

1   Complete these sentences about hormones in the human body.
Choose your words from the list.

bile   **blood**   **digestive**   **endocrine**   **neurones**   **pituitary**   **thyroid**

Hormones are produced by cells in the _____ glands.

Many of these glands are controlled by hormones from the _____ gland in the brain.

Hormones are carried to their target organs in the _____ .   **[3]**

2   This is an extract from an article published in a magazine.

> **Ricky's Story**
>
> Ricky is 16 years old and found out that he had diabetes when he was 10.
>
> 'At first it was a big change because I had to learn to do my injections and not to eat all the sweet things I used to eat. Even with my diabetes, I still play sport regularly and I'm hoping to become a PE teacher.'

a)   What type of diabetes does Ricky have?   Answer: _____   **[1]**

b)   What would Ricky be injecting into his body?   Answer: _____   **[1]**

c)   Suggest what problems there would be if Ricky did not inject himself.

_____

_____   **[2]**

d)   Ricky's uncle Gary also has diabetes.
Gary is 50 years old and has only just developed diabetes.

     i)   What type of diabetes does Gary have?   Answer: _____   **[1]**

     ii)   How would the treatment for Gary's diabetes differ from Ricky's treatment?

_____

_____   **[2]**

Total Marks _____ / 10

Hormones and Reproduction

1 The graph in **Figure 1** shows the levels of two hormones in a woman's body at different stages of the menstrual cycle.

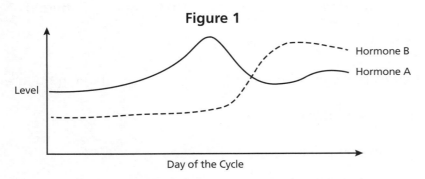

Figure 1

a) Both hormones are produced by the ovaries. Complete the following table.

	Name of Hormone	One Function of Hormone
Hormone A		
Hormone B		

[4]

b) Mark on the graph the approximate time of the following events:

    i) Mark the time of ovulation with an **X**. [1]

    ii) Mark the time of menstruation with a **Y**. [1]

c) i) What is the name of the male sex hormone?

                                     Answer: .... [1]

    ii) Where is the male sex hormone produced?

                                       Answer: .... [1]

2 Explain how the hormones in the contraceptive pill can prevent pregnancy occurring.

....

....

....

....

[3]

Total Marks .... / 11

Sexual and Asexual Reproduction

1. **Figure 1** shows a strawberry plant reproducing both sexually and asexually.

Figure 1

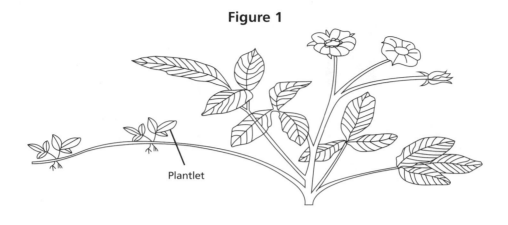

Plantlet

a) Describe how the plant reproduces asexually.

....

.... **[2]**

b) Describe how the plant reproduces sexually.

....

.... **[2]**

2. a) Complete these sentences about the genetic material in a cell.

   The genetic material in the cell is carried on long strands called ....

   These strands are found in the nucleus and are made of a chemical called ....

   A section of a strand that codes for one protein is called a .... **[3]**

   b) Write down **one** reason why it is useful for scientists to study our genetic material.

   ....

   .... **[1]**

   Total Marks .... / 8

Patterns of Inheritance

1 Gaucher disease (GD) is a genetic condition caused by a recessive allele.

Figure 1 shows part of a family tree showing some individuals that have Gaucher disease.

Figure 1

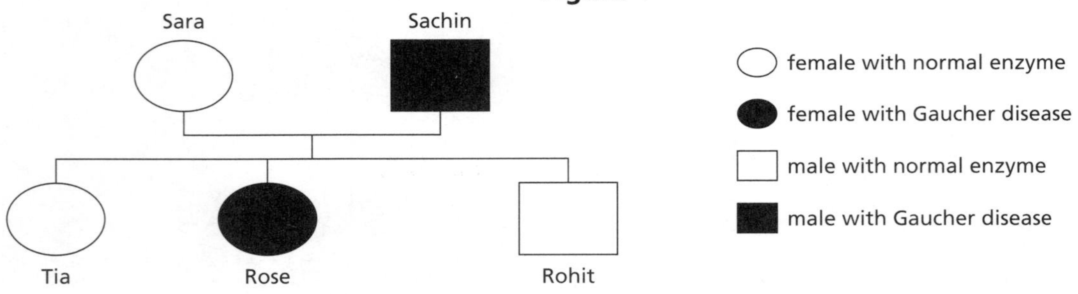

a) Write down the name of **one** person who is:

   i) **definitely** homozygous for this gene.   Answer: .... **[1]**

   ii) **definitely** heterozygous for this gene.   Answer: .... **[1]**

b) **Table 1** gives information about the different family members. Complete the table.

Table 1

Family Member	Possible Genotype	Phenotype
Sachin		
Tia	Gg	

[3]

c) Sara and Sachin are expecting another child.

   What is the probability that the child that they are expecting is a boy?

   Probability: .... **[1]**

d) A pregnant woman can have her unborn baby tested for Gaucher disease.

   Suggest why she might decide **not** to have her baby tested even if it may have Gaucher disease.

   ....

   .... **[2]**

Total Marks .... / 8

Variation and Evolution

1   Bill and Ben are identical twins.
This means that they have inherited the same genes from their parents.

Figure 1

Write each of the characteristics from **Figure 1** in the correct column of the table.

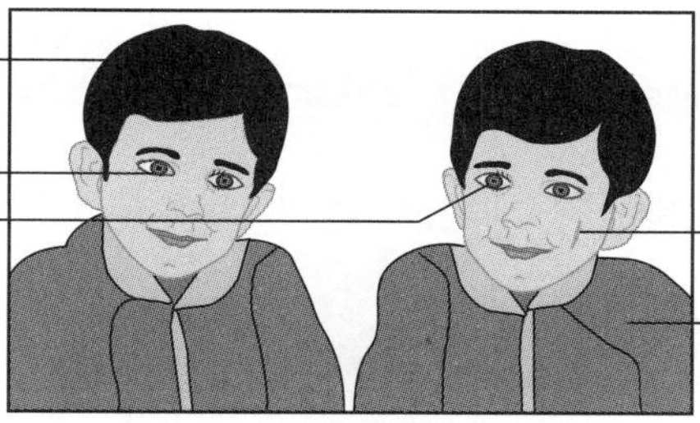

Bill is 160cm tall

Bill and Ben have brown eyes

Ben has a scar

Ben's body mass is 60kg

Controlled by Their Genes	Caused by the Environment	Controlled by their Genes and the Environment

[4]

2   **Figure 2** shows two forms of a moth that rests on trees. These moths are eaten by birds.

Figure 2

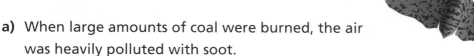

a)   When large amounts of coal were burned, the air was heavily polluted with soot.

Suggest what would happen to the colour of tree trunks in a heavily polluted area.

.... [1]

b)   A survey of dark coloured and light coloured moths was carried out in a polluted area. An equal number of light and dark moths were collected, marked and released. Several days later, moths were recaptured.

Suggest why more marked dark moths were recaptured than marked light moths.

....

....

.... [2]

c)   Over a long period of time the population of moths in the polluted area may change.

What is the name of the theory that explains how this may happen?

Answer: .... [1]

Total Marks .... / 8

Manipulating Genes

1 The following sentences are about processes that involve manipulating genes

Complete the sentences with words from the list.

genetic engineering   **selective breeding**   **natural selection**

Farmers manipulate the genes of animals, by carefully selecting which animals mate.

This is called _____ .

It is now possible to make bacteria produce human insulin by the process of _____ . **[2]**

2 Many of the varieties of strawberry grown by farmers have been produced by selective breeding.
Table 1 shows information about four varieties of strawberry.

Table 1

Variety	Month When Strawberries Ripen	Yield	Pattern of Growth
Cambridge Vigour	June	high	upright
Pegasus	July	high	spreading
Elvira	May	very high	spreading
Calypso	September	medium	bushy

a) i) Which strawberry would a farmer grow if he wanted fruit to sell as early as possible in the year?

Answer: _____ **[1]**

ii) A gardener only has a small garden.

Explain why he grows Cambridge Vigour.

_____

[1]

b) A farmer wants to produce a variety of strawberry plant that will give very high yields of strawberries in September.

i) Suggest which **two** varieties the farmer might use to obtain this new variety.

_____

[1]

ii) Describe how the farmer would produce this new variety of strawberry plant.

_____

_____

_____

[3]

Total Marks _____ / 8

Classification

1 a) What is meant by the term 'species'?

....

.... **[2]**

b) **Table 1** shows the Latin names of some different cats.

i) Two of the cats are more closely related than the others.

Write down the common names of these **two** cats.

Table 1

Common Name	Latin Name
bobcat	*Felis rufus*
cheetah	*Acinonyx jubatus*
lion	*Panthera leo*
ocelot	*Felis pardalis*

.... **[2]**

ii) Explain your answer to part **i)** above.

....

.... **[2]**

2 Below is an extract from an article about the Great Bustard.

> The Great Bustard is a massive bird. It has a wingspan of nearly two metres and used to be a great sight as it flew over the British countryside. However, in the 1870s, it became extinct in Britain. The problem was that they needed a lot of space around them to mate. People, machinery or animals nearby would disturb the birds. They were also widely hunted.

a) In the article, what is meant by the word extinct?

.... **[1]**

b) Hunting by man was partly responsible for the Great Bustard becoming extinct in Britain.

Write down **one** other way in which man contributed to the bird becoming extinct.

.... **[1]**

c) Give **one** way in which scientists can find out information about animals that became extinct a long time ago.

.... **[1]**

Total Marks .... / 9

Ecosystems

1 Camels live in desert areas of Africa.

a) Explain how the following special adaptations help the camel survive in the desert.

Figure 1

i) Webs of skin between their toes.

.... [1]

ii) A store of fat in a hump on the top of their body.

.... [1]

b) About one hundred years ago, camels were introduced into Australia.

Complete the sentences using words from the list.

abiotic   biotic   community   population   habitat   competed   reproduced

The camels that were introduced formed a large interbreeding ....

The area that they were introduced into was a new .... for them and they were well adapted to live there.

The .... made up of other plants and animals living there started to suffer.

The camels .... with cattle for .... factors such as food.   **[5]**

c) Scientists want to estimate how many organisms there are in an area of Australia.

i) Describe how they could estimate the number of cacti growing in that area.

....

....

....

....

....

.... **[6]**

ii) Why is it harder to estimate the number of camels in the area?

....

....

.... **[2]**

Total Marks .... / 15

Cycles and Feeding Relationships

1   **Figure 1** shows part of the carbon cycle.

Complete the diagram by writing the names of the correct processes in the empty boxes.

Figure 1

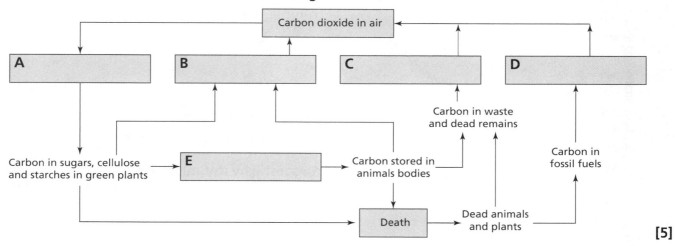

[5]

2   The graph in **Figure 2** shows how the number of rabbits and foxes in one square kilometre of woodland varies over a 40-month period.

Figure 2

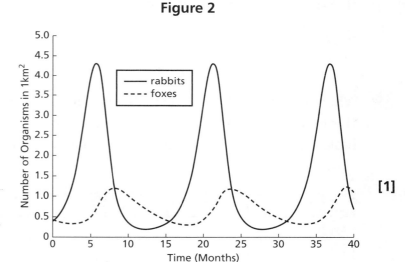

a) What name is given to this type of graph?

Answer: _____ [1]

b) If the rabbits are primary consumers in the woodland, what term is used to describe the foxes?

Answer: _____ [1]

c) Explain why the fox numbers change throughout the 40-month period.

_____

_____

_____

[3]

Total Marks _____ / 10

Disrupting Ecosystems

1 **Figure 1** shows the levels of carbon dioxide in the air above a small island in the Pacific Ocean. Scientists measured the carbon dioxide here to see if levels in the Earth's atmosphere are increasing.

Figure 1

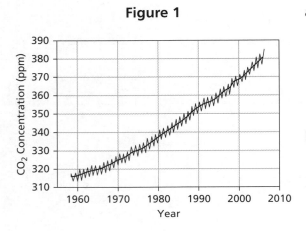

a) Suggest why scientists measured the levels at this remote island.

....

.... **[1]**

b) Levels of carbon dioxide go up slightly each winter and down each summer.

Suggest why this might be.

....

.... **[2]**

c) Explain how increasing carbon dioxide levels could lead to global warming.

....

....

.... **[3]**

2 When acid rain falls into a lake, the water becomes more acidic.
This may change the types of animals and plants that can live there.
Table 1 shows a number of aquatic animals and the range of pH in which they are found.

Table 1

Name of Animal	pH 4.0	pH 4.5	pH 5.0	pH 5.5	pH 6.0	pH 6.5
perch		✓	✓	✓	✓	✓
frog	✓	✓	✓	✓	✓	✓
clam					✓	✓
crayfish				✓	✓	✓

a) Which animal can survive the most acidic conditions?

Answer: .... **[1]**

b) Explain how acid rain is caused.

....

....

.... **[3]**

Total Marks .... / 10

Atoms, Elements, Compounds and Mixtures

1 A student heated a piece of copper in air.
The word equation for the reaction is:

copper + oxygen ➜ copper(II) oxide

a) Which equation correctly represents the reaction?
Tick **one** box.

$Cu + O → CuO$ ☐ $2Cu + O_2 → 2CuO$ ☐

$2Cu + 2O → 2CuO$ ☐ $Cu_2 + O_2 → 2CuO$ ☐ [1]

b) The mass of the copper before heating was 15.6g.
The mass of copper oxide produced was 18.9g.

Calculate the mass of oxygen that was used in the reaction.

Answer: .... g [1]

c) What was the resolution of the balance the student used to measure the masses?

Answer: .... g [1]

2 A student was asked to produce pure water from salt solution.

a) Describe how they would do
this using the equipment in **Figure 1**.

Figure 1

Condenser

Round
bottomed
flask

Bunsen
burner

Beaker

....

....

....

....

[3]

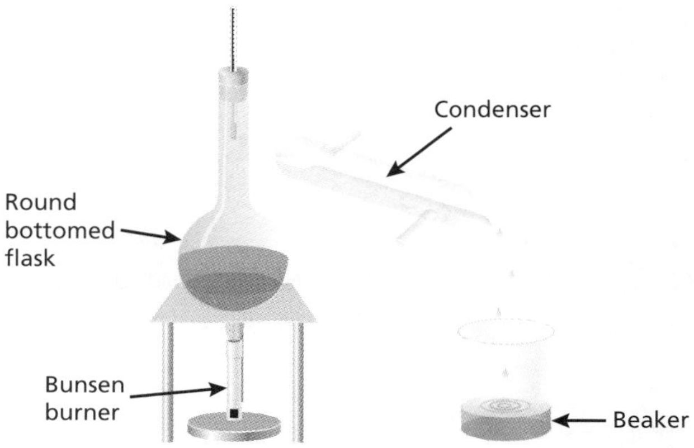

b) State **one** hazard when carrying out
this procedure.

....

[1]

Total Marks .... / 7

Atoms and the Periodic Table

1 Early models of atoms showed them as tiny spheres that could not be divided into simpler substances.

Figure 1

Sea of positive charge

In 1897, Thomson discovered that atoms contained small negatively charged particles. He proposed a new model shown in **Figure 1**.

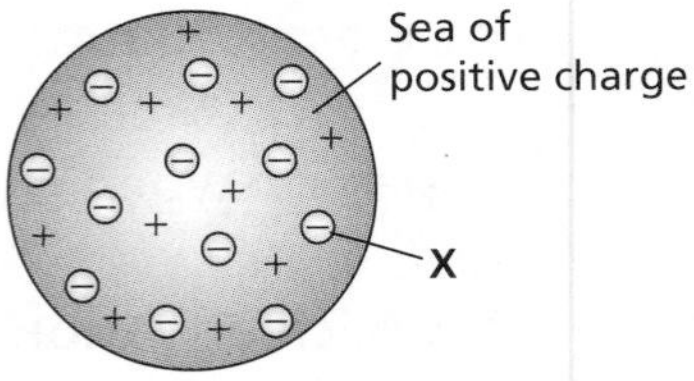

a) Why did the model of the atom have to change?

[2]

b) Name the particle labelled **X**.

Answer: [1]

2 In 1909, Geiger and Marsden bombarded a thin sheet of gold with positively charged alpha particles. They found that most passed through, but some were deflected back.

a) Explain why this result was unexpected.

[1]

b) Explain why they would have repeated their experiment several times.

[2]

c) Describe the new model of the atom that resulted from this experiment.

[3]

Total Marks / 9

The Periodic Table

1 An unknown element has the electronic configuration 2,8,8,7.

Where would it be found in the periodic table shown in **Figure 1**?

Tick **one** box.

Figure 1

A ☐   B ☐

C ☐   D ☐

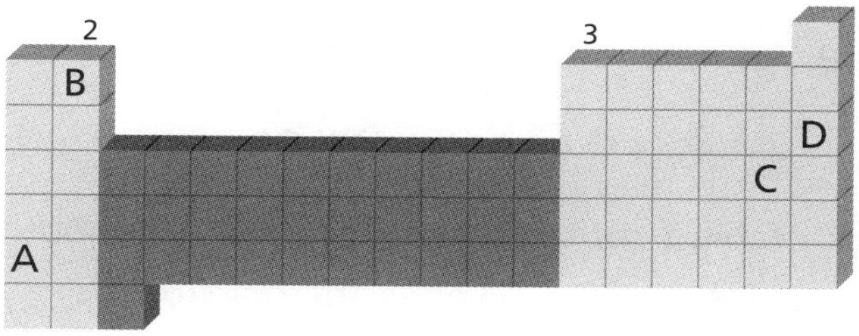

[1]

2 **Table 1** shows some data on the physical properties of elements.

Table 1

Physical Property	Element W	Element X	Element Y	Element Z
Melting Point (°C)	−38.82	−189.34	180.5	1538
Density (g/cm³)	13.53	1.40	0.53	7.87
Conductor of Electricity?	Yes	No	Yes	Yes

a) Which element, **W**, **X**, **Y** or **Z**, is a non-metal?
Give a reason for your answer.

....

....
[2]

b) Which element, **W**, **X**, **Y** or **Z**, is mercury?
Give a reason for your answer.

....

....

....
[3]

c) One of the elements is the Group 1 metal lithium.

Complete the word equation to show the reaction of lithium with water.

lithium + water ➔ lithium .... + ...   [2]

Total Marks .... / 8

States of Matter

1 A student burns magnesium.

It reacts with oxygen in the air to form a white powder called magnesium oxide.

a) Add the missing state symbols to the equation for the reaction.

$2Mg(s) + O_2($ _____ $) \rightarrow 2MgO($ _____ $)$ **[2]**

Table 1 contains some information about the melting and boiling points of the substances involved in the reaction.

Table 1

	Magnesium	Oxygen	Magnesium Oxide
Melting Point (°C)	650	−219	2830
Boiling Point (°C)	1091	−183	3600

b) State the temperature at which magnesium would change from a solid into a liquid.

Answer: _____ °C **[1]**

c) State the temperature that oxygen gas would have to be cooled to in order for it to condense.

Answer: _____ °C **[1]**

d) Explain, in terms of bonding, the difference in boiling points between oxygen and magnesium oxide.

_____

_____

_____

_____

_____

_____

_____

_____

[6]

Total Marks _____ / 10

Ionic Compounds

1 A student investigated the properties of some different compounds.

a) Which of the following compounds contain ionic bonds?
Tick **two** boxes.

Water (H_2O) ☐   Glucose ($C_6H_{12}O_6$) ☐

Calcium chloride ($CaCl_2$) ☐   Sodium carbonate (Na_2CO_3) ☐

Hydrogen chloride (HCl) ☐   **[2]**

b) The student discovered that the ionic compounds would not conduct electricity when solid but they would when dissolved in water.

Explain why.

....

....

....   **[3]**

2 **Figure 1** shows the outer electrons in an atom of the Group 2 element magnesium and in an atom of the Group 7 element bromine.
Magnesium forms an ionic compound with bromine.

Describe what happens when **one** atom of magnesium reacts with **two** atoms of bromine.
Give your answer in terms of electron transfer.
Give the formulae of the ions formed.

Figure 1

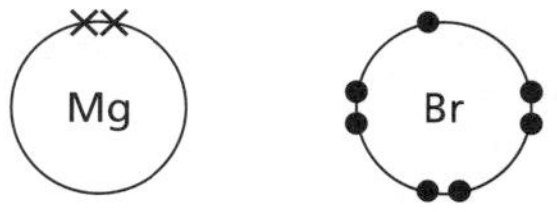

....

....

....

....

....   **[5]**

Total Marks .... / 10

Covalent Compounds

1 Fluorine and bromine are elements found in Group 7 of the periodic table.
At room temperature, fluorine is a gas and bromine is a liquid.

Why, at room temperature, is fluorine a gas and bromine a liquid?
Tick **one** box.

The covalent bonds between bromine are stronger. ☐

Bromine has a giant covalent structure and fluorine is a simple molecule. ☐

The forces between bromine molecules are stronger. ☐

Fluorine contains fewer molecules than bromine. ☐ [1]

2 Bromine reacts with hydrogen to form hydrogen bromide.

Figure 1

a) Complete the dot and cross diagram in **Figure 1** to show
the covalent bonding in a molecule of hydrogen bromide.

Show the outer shell electrons only.

[2]

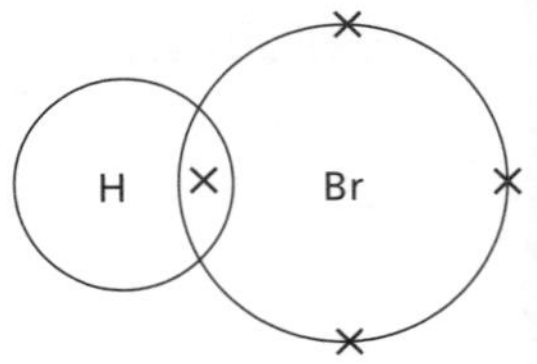

b) State the formula for a molecule of hydrogen bromide.

Answer: _____ [1]

3 Diamond has a giant covalent structure made up of carbon atoms.
It has a high melting point and does not conduct electricity when molten.

Draw **one** line from each property to the explanation of that property.

Property	Explanation of Property
	Strong covalent bonds between many carbon atoms
High melting point	Atoms are free to move
	Weak bonds between carbon atoms
Does not conduct electricity when molten	There are no charged particles that are free to move
	Carbon atoms are in a regular arrangement

[2]

Total Marks _____ / 6

Metals and Special Materials

1   **Figure 1** shows the bonding in the metal copper.

Figure 1

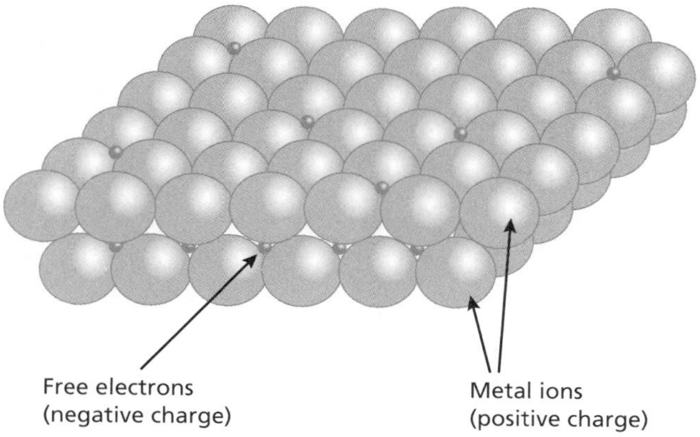

Free electrons
(negative charge)

Metal ions
(positive charge)

Copper is useful as a material for making saucepans.
This is because it has the following properties:

- high melting point
- good thermal conductor
- malleable (can be easily shaped).

Explain, in terms of its metallic bonding, why copper has these properties.

....

....

....

....

....

....

....

....

[6]

Total Marks .... / 6

Conservation of Mass

1 A student was asked to carry out the reaction shown in the word equation:

zinc + hydrochloric acid → zinc chloride + hydrogen

a) What is the balanced symbol equation for the reaction?
Tick **one** box.

$Zn(s) + 2HCl(aq) \rightarrow ZnCl(aq) + H(g)$ ☐

$Zn(s) + HCl(aq) \rightarrow ZnCl_2(aq) + H_2(g)$ ☐

$Zn(s) + 2HCl(aq) \rightarrow ZnCl_2(aq) + H_2(g)$ ☐

$Zn(s) + 2HCl(aq) \rightarrow ZnCl_2(aq) + H(g)$ ☐ **[1]**

b) The student plans to use the equipment in **Figure 1**.
The teacher tells the student not to use the bung.

Explain why.

Figure 1

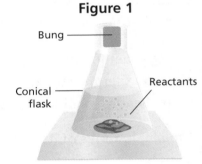

Bung

Conical flask

Reactants

....

....

....

....

[2]

c) The student carries out the reaction on a balance, as shown in **Figure 2**.

Figure 2

Explain what happens to the mass reading during the reaction.

....

....

....

....

....

[3]

Total Marks .... / 6

Reactivity of Metals

1 This question is about how metals are extracted from their ores.
Figure 1 shows the reactivity series of metals.

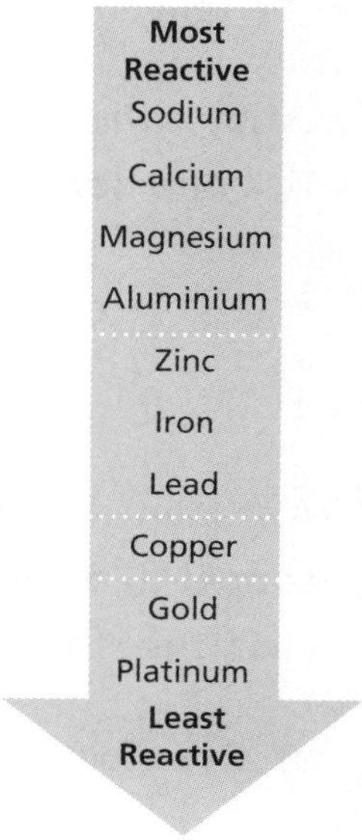

a) Zinc can be found naturally as the compound zinc(II) oxide.

Use the information in **Figure 1** to decide which of these metals can be used to displace zinc from zinc(II) oxide.
Tick **two** boxes.

Magnesium ☐

Iron ☐

Copper ☐

Sodium ☐

Figure 1

Most Reactive
Sodium
Calcium
Magnesium
Aluminium
Zinc
Iron
Lead
Copper
Gold
Platinum
Least Reactive

[2]

b) In reality, carbon is used to extract zinc from zinc(II) oxide.

Suggest **one** reason why carbon is used rather than another metal.

....

.... [1]

2 A student reacted zinc with a solution of copper chloride.
The equation for the reaction is:

zinc + copper chloride → zinc chloride + copper

What kind of reaction is this?
Tick **one** box.

Neutralisation ☐

Thermal decomposition ☐

Displacement ☐

Combustion ☐

[1]

Total Marks .... / 4

The pH Scale and Salts

1 Universal indicator was added to a sample of an unknown solution.
The universal indicator turned orange.

What is the pH value of the solution?
Tick **one** box.

1 ☐   7 ☐

4 ☐   10 ☐   **[1]**

2 A student was asked to prepare pure, dry crystals of a soluble salt using insoluble copper(II) oxide and hydrochloric acid.
First, the student measured out 25cm³ of acid into a conical flask.

a) Describe how the student should complete the preparation of the salt.

_____

_____

_____

_____

_____

_____

_____

_____   **[4]**

b) State **one** safety precaution that the student should take during the preparation.

_____

_____   **[1]**

c) Name the salt produced.

Answer: _____   **[1]**

Total Marks _____ / 7

Electrolysis

1 Aluminium is extracted from its ore, aluminium oxide, using electrolysis, as shown in **Figure 1**.

Figure 1

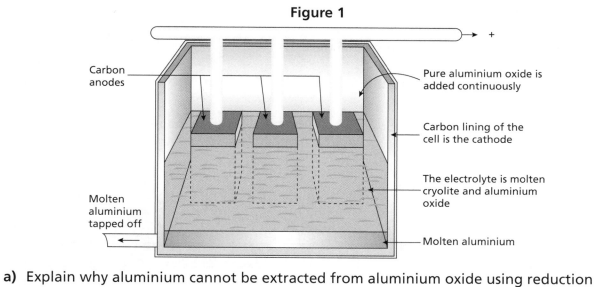

Carbon anodes

Pure aluminium oxide is added continuously

Carbon lining of the cell is the cathode

The electrolyte is molten cryolite and aluminium oxide

Molten aluminium tapped off

Molten aluminium

a) Explain why aluminium cannot be extracted from aluminium oxide using reduction with carbon.

....

.... **[1]**

b) State the product formed at the:

i) anode: .... ii) cathode: .... **[2]**

c) Explain why the extraction of aluminium from aluminium oxide is an expensive process and describe the methods used by manufacturers to reduce the cost.

....

....

....

....

....

....

.... **[3]**

Total Marks .... / 6

Exothermic and Endothermic Reactions

1 A student carried out an investigation into how the reactivity of metals affects how exothermic or endothermic their reaction with dilute hydrochloric acid is.

The student used the apparatus shown in **Figure 1**.

The student's results are shown in **Table 1**.

Figure 1

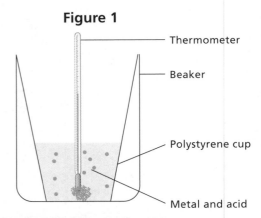

Thermometer

Beaker

Polystyrene cup

Metal and acid

Table 1

Metal	Temperature at Start (°C)	Highest Temperature Reached (°C)
Zinc	21.0	30.1
Copper	21.2	21.3
Magnesium	21.4	82.6
Iron	21.4	26.0

a) Describe the function of the polystyrene cup.

....

....

.... **[2]**

b) State **two** control variables the student should have used in this investigation.

....

.... **[2]**

c) The order of reactivity of these metals from highest to lowest is: magnesium, zinc, iron, copper.

Write a conclusion that answers the original question that the student was investigating.

....

....

.... **[2]**

Total Marks .... / 6

Rate of Reaction

1 When hydrochloric acid reacts with sodium thiosulfate one of the products is sulfur, which is insoluble.

A student carried out an investigation into the following hypothesis:

As the concentration of acid increases, the rate of the reaction will increase.

The student added sodium thiosulfate to dilute hydrochloric acid of concentration 0.25mol/dm³ in a conical flask.
The conical flask was placed on a cross, as shown in **Figure 1**.
The student timed how long it took before they could no longer see the cross.

Figure 1

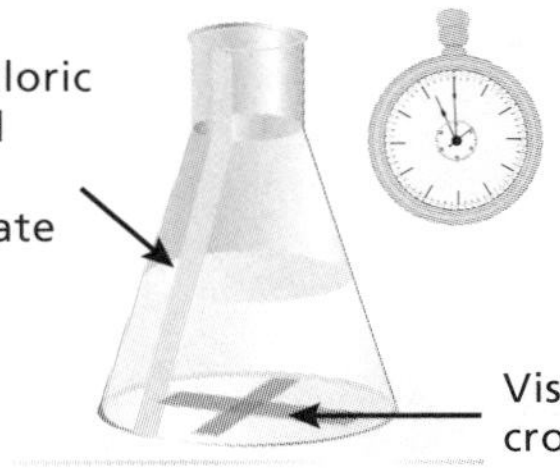

Hydrochloric acid and sodium thiosulfate

Visible cross

a) Describe what the student should do next in order to investigate the hypothesis.
   You should include a suitable range for the independent variable in your answer.

   ....

   ....

   ....

   .... **[2]**

b) State **one** safety precaution the student should take.

   .... **[1]**

c) Use collision theory to write a prediction for what will happen if the hypothesis is correct.

   ....

   ....

   ....

   ....

   ....

   .... **[4]**

Total Marks .... / 7

Reversible Reactions

1 This question is about the reaction between ammonia and hydrogen chloride.

The equation for the reaction is: $NH_3(g) + HCl(g) \rightleftharpoons NH_4Cl(s)$

The reaction can be carried out using the apparatus shown in **Figure 1**.

Figure 1

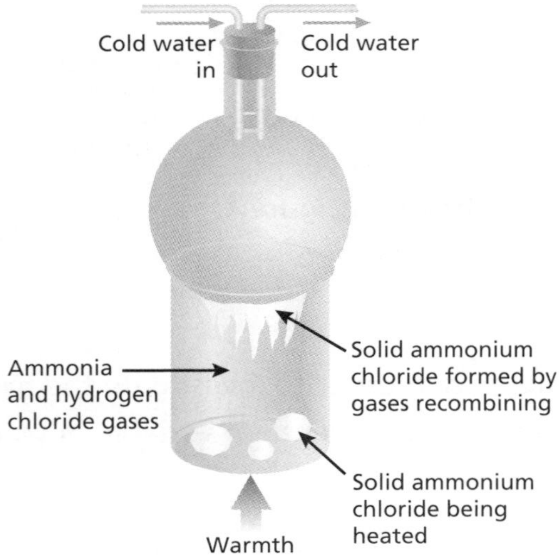

a) What is meant by the symbol $\rightleftharpoons$ in the equation?

Answer: _____ **[1]**

b) Which reaction, the **forward reaction** or the **reverse reaction**, is exothermic?
Give a reason for your answer.

_____

_____

_____ **[3]**

c) The reaction taking place in **Figure 1** is at equilibrium.

What does this mean?
Tick **two** boxes.

Only ammonium chloride is being produced. ☐

The forward and reverse reactions are taking place at the same rate. ☐

The amounts of reactants and products are constant. ☐

The reaction has completed. ☐ **[2]**

Total Marks _____ / 6

Alkanes

1 Fractional distillation is used to separate crude oil into useful mixtures called fractions.
It takes place in a column, as shown in **Figure 1**.

a) The molecules found in crude oil are mostly hydrocarbons.

What is a hydrocarbon?

....

....

....

.... [2]

Figure 1

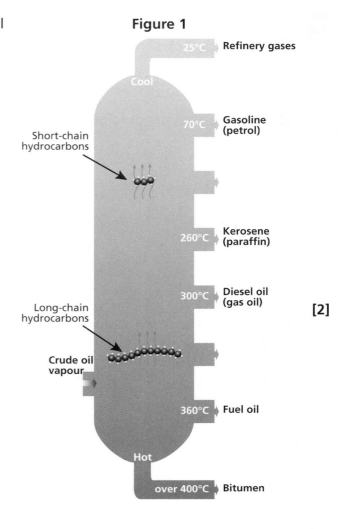

b) This is what takes place inside the column:

1. Crude oil is heated until it forms a vapour.
2. The vapour rises up the tower.

Describe what happens next in order for the fractions to form.

....

....

....

....

....

.... [2]

2 Methane (CH_4) is the name of a hydrocarbon found in refinery gas.

Complete the balanced symbol equation for the complete combustion of methane.

CH_4 + 2 .... → CO_2 + ... H_2O [2]

Total Marks .... / 6

Cracking Hydrocarbons

1 Hydrocarbon molecules can be cracked to form molecules with shorter chains.
The equation shows the cracking of $C_{20}H_{42}$.

$$C_{20}H_{42} \rightarrow C_8H_{18} + \boxed{} C_5H_{10} + C_2H_4$$

a) What number needs to go in the box to balance the equation?
Tick **one** box.

1 ☐

2 ☐

3 ☐

4 ☐   [1]

b) Which products are alkanes?
Tick **one** box.

C_8H_{18} and C_2H_4 ☐

C_5H_{10} and C_2H_4 ☐

C_8H_{18} only ☐

C_8H_{18}, C_5H_{10} and C_2H_4 ☐   [1]

2 A student is given a test tube of colourless liquid hydrocarbon.

Describe a test the student could carry out to find out if the liquid is an alkane or an alkene.
You should state the chemical used for the test and the results of the test for both an alkane and an alkene.

....

....

....

....

....

....

[3]

Total Marks .... / 5

Chemical Analysis

1 A student was asked to investigate five different inks using chromatography.
Figure 1 shows the apparatus they needed to use.
The student set up the apparatus correctly and left it for several minutes.
Figure 2 shows their results.

Figure 1

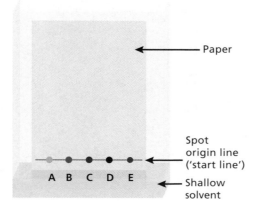

Figure 2

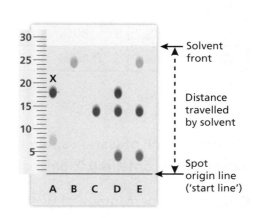

a) Which part of the apparatus is the stationary phase?

Answer: _____ **[1]**

b) The student was not sure whether to use a pen or pencil to draw the origin line.

State which they should choose. Give a reason for your answer.

_____

_____ **[2]**

c) Which **two** inks are pure? Give a reason for your answer.

_____

_____ **[2]**

d) Calculate the R_f value of **X**.
Give your answer to 2 decimal places.

Answer: _____ **[3]**

Total Marks _____ / 8

The Earth's Atmosphere

1 **Table 1** shows some of the gases that are in the Earth's atmosphere today.

Table 1

	Oxygen	Nitrogen	Carbon Dioxide
A	80%	20%	0.04%
B	0.04%	80%	20%
C	20%	80%	0.04%
D	20%	0.04%	80%

Which letter, **A**, **B**, **C** or **D**, represents the correct proportions?
Tick **one** box.

A ☐

B ☐

C ☐

D ☐ **[1]**

2 The percentage of water vapour in the atmosphere can vary.
$4.2dm^3$ of air contains $0.05dm^3$ of water vapour.

Calculate the percentage of water vapour in the air.
Give your answer to 2 significant figures.

Answer: .... % **[2]**

3 One theory for how the Earth's early atmosphere was formed is that gases were released by volcanoes.
Scientists have recently proposed a new theory: that the early atmosphere was formed by comets hitting the Earth.

a) What would the scientists need for this new theory to be accepted?

....

.... **[2]**

b) Explain why we cannot be sure if either theory is correct.

....

....

....

.... **[3]**

Total Marks .... / 8

Greenhouse Gases

1 Many scientists believe that human activities are causing the mean global temperature of the Earth to increase.

They think that this is because of an increase in the amounts of greenhouse gases in the atmosphere.

Water vapour is an example of a greenhouse gas.

a) Name one other greenhouse gas and state how human activity has increased its amount in the atmosphere.

....

....

.... **[2]**

In 2000, computer models were used to predict how the mean global temperature might change in the future.

Figure 1 shows the results. Each line shows the predictions made using a different model.

Figure 1

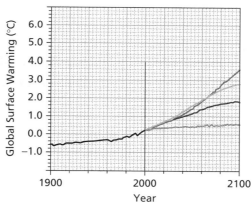

b) Calculate the range of global surface warming as predicted for 2100 by the different models.

From: .... To: .... **[2]**

c) Explain why it is difficult to produce models to predict future climate change.

....

....

.... **[2]**

Total Marks .... / 6

Earth's Resources

1 In the UK, potable water is produced from an unpolluted source of fresh water.
Potable water contains low levels of dissolved substances.

a) What is potable water?

.... [1]

b) Explain why potable water cannot be called pure in the chemical sense.

....

....

.... [2]

The stages of producing potable water from fresh water are:

1. Pass the water through filter beds to remove any solids.
2. Sterilise the water to kill microorganisms.

c) Give **one** way of sterilising the water.

.... [1]

In other parts of the world, sea water is used as a source of potable water.
To produce potable water from sea water, a process called distillation is used.

d) Explain why distillation is used.

....

.... [1]

e) Explain why producing potable water from sea water is more expensive than producing
it from fresh water sources.

....

....

.... [2]

Total Marks _____ / 7

Using Resources

1. Shopping bags can be made out of paper or plastic.
   Table 1 is part of a LCA (life cycle assessment) comparing these two materials.

Table 1

	Paper Bag	Plastic Bag
Raw Materials	Wood pulp	Crude oil
Manufacture	• Wood pulp is added to water (up to 100 times the mass of the pulp) and mixed. • Clay, chalk or titanium oxide is added. • The mixture is squeezed and heated to remove the water.	• Crude oil is heated to 360°C during fractional distillation. • A fraction is cracked at 850°C to produce alkenes. • Polymerisation of alkenes at 150°C.
Transport	• Average mass = 55g • Seven trucks needed to transport two million bags.	• Average mass = 7g • One truck needed to transport two million bags.
Use During Lifetime	• Not normally reused.	• Can be reused many times.
Disposal at the End of Life	• Taken to landfill (biodegradable). • Can be incinerated (burned). • Can be recycled.	• Taken to landfill (not biodegradable). • Can be incinerated (burned). • Difficult to recycle.

Use the information in **Table 1** to compare the **advantages** and **disadvantages** of both types of bag.

....

....

....

....

....

[6]

Total Marks .... / 6

Forces – An Introduction

1 When a plane is in flight, the engines provide a thrust force that pushes the aircraft forwards. The wings provide a 'lift' force that acts upwards.

a) Name **two** other forces that act on the plane.
In each case state whether it is a **contact** force or a **non-contact** force.

_____

_____ **[4]**

b) The plane has a mass of 120 000kg. Calculate the weight of the plane (g = 10N/kg).

Weight = _____ N **[2]**

c) When flying at a constant altitude, how will the lift force provided by the wings compare to the weight of the airplane?
Tick **one** box.

The lift force will be less than the weight. ☐

The lift force will be the same as the weight. ☐

The lift force will be more than the weight. ☐ **[1]**

d) At times during the flight, the force from the engines is exactly balanced by the resistive forces.

How can the speed be described at this time?

_____ **[1]**

2 A student carries out an investigation into forces.
They use an air track, which suspends a glider vehicle on a cushion of air so that it can move smoothly without touching the ground.

a) What force is the air track designed to reduce?

Answer: _____ **[1]**

b) Use the idea of contact and non-contact forces to explain why the air track is effective at doing this.

_____

_____ **[2]**

Forces in Action

1 A student carries out an investigation involving springs.
   The student suspends a spring from a rod.
   A force is applied to the spring by hanging a mass from it.

   a) Describe how the student could test if the spring is behaving elastically.

   ....

   .... [2]

   b) In a second investigation, the student takes a set of measurements for force and extension.
      The results are shown in **Table 1** below.

   Table 1

Force (N)	0.0	1.0	2.0	3.0	4.0	5.0	6.0
Extension (cm)	0.0	4.0		12.0	16.0	22.0	31.0

   i) Add the missing value to the table. [1]

   ii) Explain why you chose this value.

   ....

   .... [2]

   c) Complete the following sentences about the experiment in part **b)**.

   The independent variable investigated was the ....

   It had a range from .... to .... [3]

   d) The student repeats the experiment in part **b)**.

   Give **two** reasons why repeating an experiment can improve the accuracy of the results.

   ....

   .... [2]

   Total Marks .... / 10

Forces and Motion

1 A person takes their dog for a walk.
The graph in **Figure 1** shows how the distance from their home changes with time.

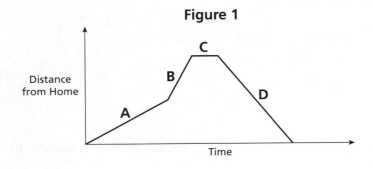

Figure 1

a) Work out the final displacement at the end of their journey.

Answer: _____ **[1]**

b) Which part of the graph, **A**, **B**, **C** or **D**, shows them walking at the fastest rate?

Answer: _____ **[1]**

c) Describe their motion during section **C**.   Answer: _____ **[1]**

d) Describe how the velocity of section **A** compares to the velocity of section **D**.

_____

_____

_____
[3]

2 Drivers on a racetrack enter a hairpin bend travelling east.
The tight bend forces them to slow down.
When they exit the bend, they are travelling west and speed back up again.
The bend is 180m long.

a) If it takes 6 seconds to travel around the bend, what is the average speed of the car around the bend?

Speed = _____ m/s **[2]**

b) A driver enters the bend at 50m/s and exits the bend at 40m/s.

i) Work out the change in speed.

Change in speed = _____ m/s **[1]**

ii) Work out the change in velocity.

Change in velocity = _____ m/s **[1]**

Total Marks _____ / 10

Forces and Acceleration

1 An experiment is carried out to investigate how changing the mass affects the acceleration of a system.
A trolley is placed on a bench and is made to accelerate by applying a constant force using hanging masses. Different masses were then added to the trolley.

Figure 1

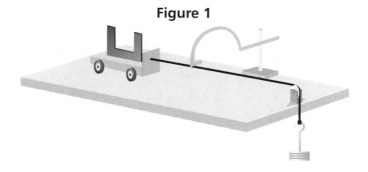

a) What is the independent variable?

Answer: _____ [1]

b) What is the control variable?

Answer: _____ [1]

c) What is the dependent variable?

Answer: _____ [1]

2 A boat accelerates at a constant rate in a straight line.
This causes the velocity to increase from 4.0m/s to 16.0m/s in 8.0s.

a) Calculate the acceleration.
Give the unit.

Answer: _____ [3]

b) A water skier being pulled by the boat has a mass of 68kg.

Use your answer to part **a)** to calculate the resultant force acting on the water skier whilst accelerating.

Answer: _____ [2]

3 The manufacturer of a car gives the following information in a brochure:
The mass of the car is 950kg.
The car will accelerate from 0 to 33m/s in 11s.

a) Calculate the acceleration of the car during the 11s.

Answer: _____ [2]

b) Calculate the force needed to produce this acceleration.

Answer: _____ [2]

Total Marks _____ / 12

Terminal Velocity, Stopping and Braking

1 The terminal velocity is the maximum velocity a falling object can reach.

Describe how a skydiver could increase their terminal velocity.

_____

_____ **[2]**

2 The graph in **Figure 1** shows how the velocity of a car changes from the moment the driver sees an obstacle blocking the road.

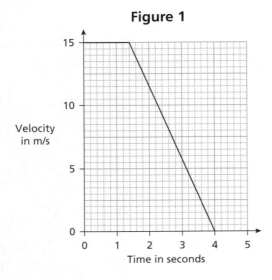

Figure 1

Velocity in m/s

Time in seconds

a) Work out the reaction time of the driver.

Answer: _____ **[1]**

b) Use the graph and your answer to part **a)** to calculate the thinking distance.

Answer: _____ **[2]**

c) Use the graph to work out the time it took for the car to stop from the moment the brakes were applied.

Answer: _____ **[1]**

d) How would the time it took for the car to stop be different if the road was wet or icy?

_____ **[1]**

e) The driver of the car was tired and had been drinking alcohol.

On the graph in **Figure 1**, sketch a second line to show how the graph would be different if the driver had been wide awake and fully alert. **[3]**

f) How would the graph look different if the vehicle had old / worn brakes and tyres?

_____

_____ **[2]**

Total Marks _____ / 12

Energy Stores and Transfers

1 An electric kettle is used to heat 2kg of water from 20°C to 100°C.

> **change in thermal energy = mass × specific heat capacity × temperature change**
> Specific heat capacity of water = 4200J/kg °C

a) All of the energy supplied to the kettle goes into the water.

Calculate the amount of electrical energy supplied to the kettle.

Answer: _____ **[3]**

b) On a different occasion, the kettle is filled with 2.5kg of water but is switched on for the same amount of time.

Use your answer to part **a)** to calculate what temperature the kettle heats the water to in this period.

Answer: _____ **[3]**

2 **Figure 1** shows a pendulum swinging backwards and forwards.
The pendulum is made from a 100g mass suspended by a light string.

Figure 1

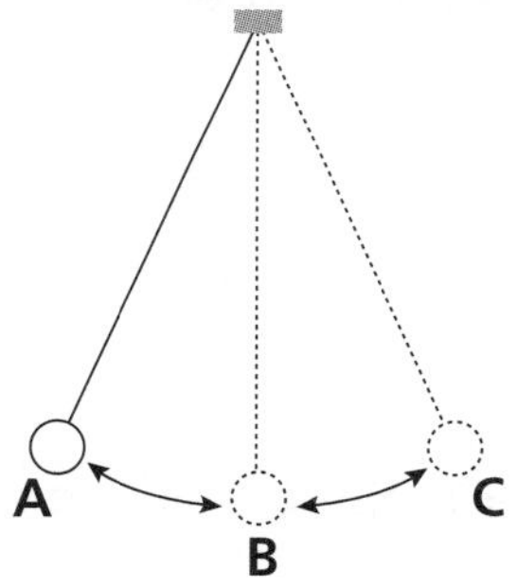

a) At position **C**, the mass is 5cm higher than at position **B**.

Calculate the difference in gravitational energy between positions **B** and **C**.
The gravitational field strength is = 10N/kg.

Answer: _____ **[3]**

b) The difference in potential energy is the same as the amount of kinetic energy gained by the mass as it swings down from **C** to **B**.

Calculate the velocity of the mass at position **B**.

Answer: _____ **[3]**

Total Marks _____ / 12

Energy Transfers and Resources

1 Complete the sentences below to explain the energy transfers involved in a solar panel.

In a solar panel, _____ energy is converted into _____ energy.

Some energy is converted into _____ energy and lost to the surroundings. **[3]**

2 A student tested four different types of fleece, **J**, **K**, **L** and **M**, to find out which would make the warmest jacket.
Each type of fleece was wrapped around a can.
The can was then filled with hot water.
The temperature of the water was recorded every 2 minutes for a 20-minute period.

The graph in **Figure 2** shows the student's results.

Figure 1

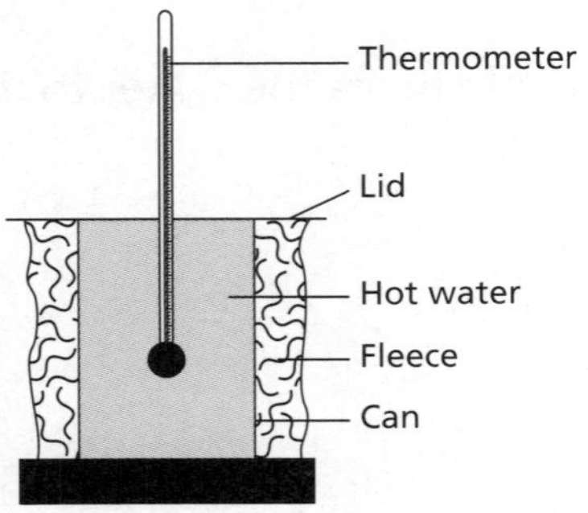

Thermometer
Lid
Hot water
Fleece
Can

Figure 2

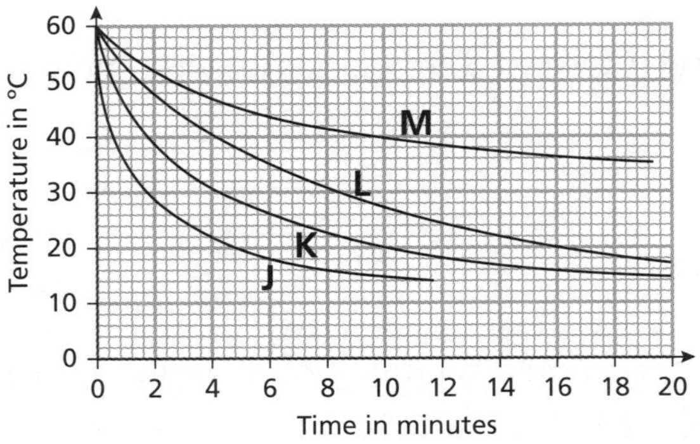

a) To be able to compare the results, it was important to use the same volume of water in each test.

Give **two** other variables that should have been kept the same in each test.

_____ **[2]**

b) Which type of fleece, **J**, **K**, **L** or **M**, should the student recommend for making a ski jacket?
You must explain your answer.

_____

_____ **[2]**

Total Marks _____ / 7

Waves and Wave Properties

1 A wave machine in a swimming pool generates waves with a frequency of 0.5Hz.

a) What does a frequency of 0.5Hz mean?

.... **[1]**

b) Give the equation that links the frequency, speed and wavelength of a wave.

.... **[1]**

c) The swimming pool is 50m long.
It takes each wave 10 seconds to travel the length of the pool.

Calculate the wave speed.

Answer: .... **[2]**

d) Use your answers to parts **b)** and **c)** to calculate the wavelength of the waves.

Answer: .... **[2]**

e) A parent notices that the waves change direction when they enter a shallow area of the pool.

What name is given to this effect?

Answer: .... **[1]**

2 Waves may be longitudinal or transverse.

Describe the differences between longitudinal and transverse waves.

....

....

....

....

.... **[3]**

Total Marks .... / 10

Electromagnetic Waves

1 a) **Figure 1** shows two beakers.
   Each beaker has a drawing pin inside.

   The first beaker is empty and the eye cannot see the drawing pin.
   The second beaker is full of water and the drawing pin can be seen.

   Explain how this is possible.

   You may add a ray line to the diagram to help with your answer.

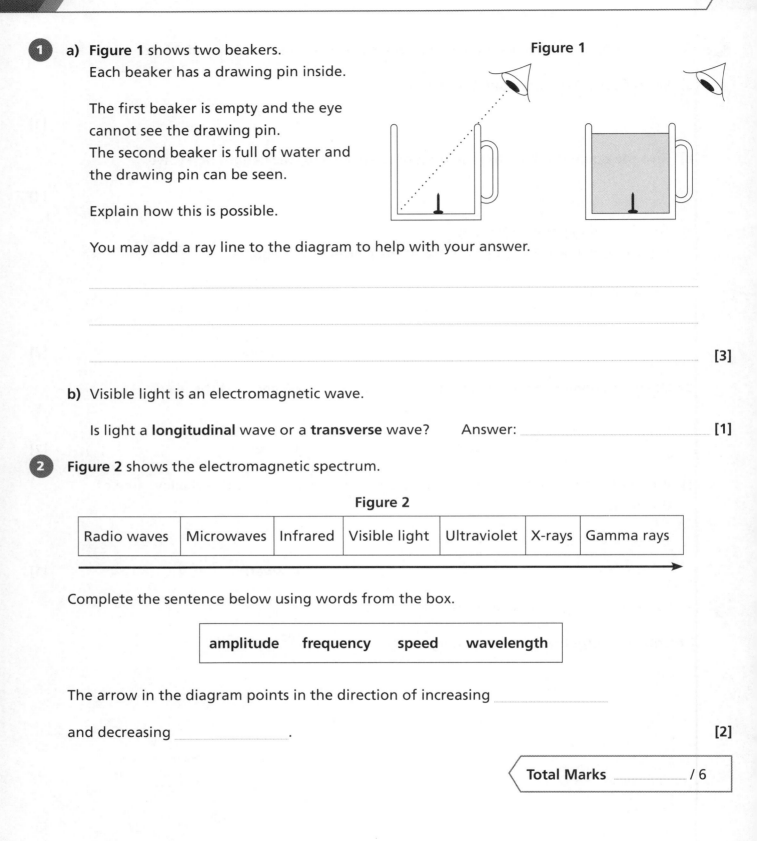

Figure 1

....

....

.... **[3]**

   b) Visible light is an electromagnetic wave.

   Is light a **longitudinal** wave or a **transverse** wave?   Answer: .... **[1]**

2 **Figure 2** shows the electromagnetic spectrum.

Figure 2

Radio waves	Microwaves	Infrared	Visible light	Ultraviolet	X-rays	Gamma rays

Complete the sentence below using words from the box.

amplitude   frequency   speed   wavelength

The arrow in the diagram points in the direction of increasing ....

and decreasing .... **[2]**

Total Marks .... / 6

The Electromagnetic Spectrum

1 Radio waves and visible light are electromagnetic waves that are used for communication.

a) Name another type of electromagnetic wave that is used for communication.

Answer: .... **[1]**

b) Name an electromagnetic wave which is **not** used for communication and give one of its uses.

....

.... **[2]**

2 After a person is injured, a doctor will sometimes request a photograph of the patient's bones

a) Which type of electromagnetic radiation would be used to produce the photograph?

Answer: .... **[1]**

b) What properties of this radiation enable it to be used to photograph bones?

....

....

.... **[2]**

3 Ultraviolet light is a type of electromagnetic wave. It is used in sunbeds.

a) State **one** hazard of sunbed use.

.... **[1]**

b) Explain why ultraviolet light is more hazardous than visible light.

....

.... **[2]**

c) Despite the risks many people still regularly use sunbeds.

Suggest **two** reasons why.

....

.... **[2]**

Total Marks .... / 11

An Introduction to Electricity

1 Circle the correct words to complete the sentences.

Electric **current** / **charge** is the flow of electrical **charge** / **potential**.

The **greater** / **smaller** the flow, the higher the **current** / **voltage**. [4]

2 The element in a set of hair straighteners has a 5A current running through it and a 230V potential difference across it.

a) Write down the equation that links potential difference, current and resistance.

Answer _____ [1]

b) Calculate the resistance of the element.

Answer _____ [2]

c) Write down the equation that links charge, current and time.

Answer _____ [1]

d) The straighteners are used for 2 minutes.

Calculate the charge that flows in this time.

Answer _____ [2]

3 Circle the correct words to complete the sentences.

Potential difference determines the amount of **energy** / **power** transferred by the charge as it passes through a component.

The **greater** / **smaller** the potential difference, the higher the **current** / **voltage** that will flow. [3]

4 Write the name of the component that each circuit symbol represents.

a) —o⁄o— Answer _____ [1]

b) —|ⱶ---|ⱶ— Answer _____ [1]

c) —▭— Answer _____ [1]

d) ⌖⊝— Answer _____ [1]

Total Marks _____ / 17

Circuits and Resistance

1 The circuit in **Figure 1** is used to measure the current and potential difference of various components.

Figure 1

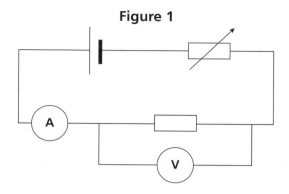

a) What is the purpose of the variable resistor?

....

.... **[2]**

b) Is the ammeter in **Figure 1** connected in series or parallel with the variable resistor?

Answer .... **[1]**

c) Is the voltmeter in **Figure 1** connected in series or parallel with the resistor?

Answer .... **[1]**

2 Draw **one** line from each component to the correct description.

Light dependent resistor (LDR)	Resistance decreases as temperature increases.
Thermistor	Resistance increases as the temperature increases.
Diode	Resistance decreases as light intensity increases.
Filament light	Has a very high resistance in one direction.

[3]

Total Marks .... / 7

Circuits and Power

1 This question is about a hairdryer that heats air and blows it out the front through a nozzle.

a) The hairdryer has an input power of 1600W.

If a person takes two minutes to dry their hair, how much energy has been transferred?

Answer _____ [3]

b) The hairdryer is 90% efficient. The remaining energy is output as sound.

How much sound energy is produced in two minutes of use?

Answer _____ [2]

2 **Figure 1** shows a series circuit with two resistors, **X** and **Y**.

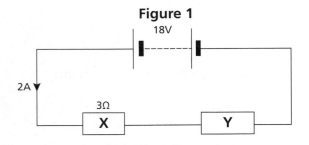

Figure 1

18V

2A

3Ω

X   Y

a) Calculate the potential difference across resistor **X**.

Answer _____ [2]

b) Use your answer to part **a)** to work out the potential difference across component **Y**.

Answer _____ [2]

c) Calculate the total resistance of the circuit.

Answer _____ [2]

Total Marks _____ / 11

Domestic Uses of Electricity

1 A battery is connected to an oscilloscope and the trace in **Figure 1** is produced.

Figure 1

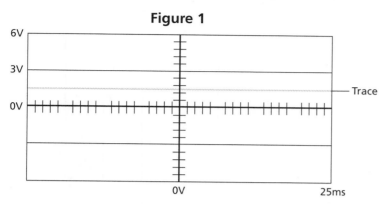

a) Use the trace to determine the potential difference of the battery and the type of current.

.... **[2]**

b) The battery is replaced by the mains supply and the trace in **Figure 2** is recorded by the oscilloscope.

Figure 2

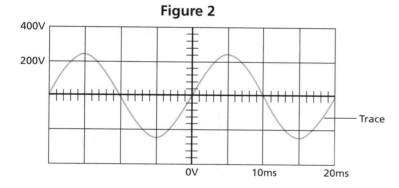

Use the trace to determine the potential difference and the type of current.

.... **[2]**

2 An appliance is switched off but the power cable is connected to the mains.

Explain how a live wire can still be dangerous.

....

....

....

.... **[4]**

Total Marks .... / 8

Electrical Energy in Devices

1 An electric blender transfers electrical energy into kinetic, heat and sound energy.

a) What is the useful energy output?   Answer _____ **[1]**

b) What happens to the waste energy produced?

_____

_____ **[2]**

2 **Table 1** gives some information about an electric drill.

Table 1

Energy Input	
Useful Energy Output	
Wasted Energy	Heat and Sound
Power Rating	500W

a) Complete **Table 1** by adding the missing types of energy.   **[2]**

b) Which of the following statements about the energy from the drill is **incorrect**?
Tick **one** box.

It spreads out and becomes more difficult to use.   ☐

It disappears.   ☐

It makes the surroundings warmer.   ☐   **[1]**

c) How much energy does the drill use per second?   Answer _____ **[1]**

3 The National Grid distributes electricity from power stations to consumers.
The voltage across the overhead cables of the National Grid is much higher than the output voltage from the power station generators.

Explain how this achieved and why it is important.

_____

_____

_____

_____

_____ **[4]**

Total Marks _____ / 11

Magnetism and Electromagnetism

1 The full name for the north pole of a magnet is the 'north-seeking pole'.

Explain what is meant by this.

....

.... **[2]**

2 The north pole of a permanent magnet is moved close to the north pole of another permanent magnet.

a) What would you expect to happen?

.... **[1]**

b) A piece of iron is moved close to the north pole of a permanent magnet.

What would you expect to happen?

.... **[1]**

3 **Figure 1** shows an electric bell.

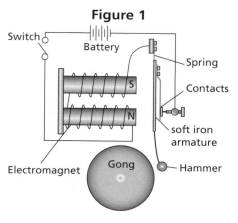

Figure 1

Describe what happens when the switch is pressed.

....

....

....

....

.... **[5]**

Total Marks _____ / 9

Particle Model of Matter

1 Heating a substance can cause it to change state from a solid to a liquid or from a liquid to a gas.

a) What is meant by 'specific latent heat of fusion'?

....

.... **[2]**

b) While a kettle boils, 0.012kg of water changes to steam.

Calculate the amount of energy required for this change.
Use the correct equation from the Physics Equation Sheet on page 167.
Specific latent heat of vaporisation of water = 2.3×10^6 J/kg

Answer .... **[2]**

2 The graph in **Figure 1** shows how temperature varies with time as a substance cools.
The graph is **not** drawn to scale.

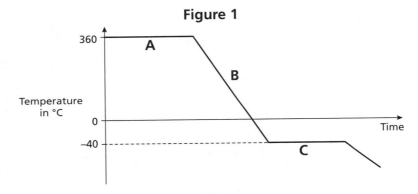

Figure 1

a) Explain what is happening to the substance in section **A** of the graph.

....

.... **[2]**

b) Explain what is happening to the substance in section **B** of the graph.

....

....

.... **[2]**

Total Marks .... / 8

Atoms and Isotopes

1   Atoms contain three types of particle.

a) Complete the table to show the relative charges of the subatomic particles.

Particle	Relative Charge
Electron	
Neutron	
Proton	

[3]

b) A neutral atom has no overall charge.

Explain why in terms of its particles.

....

....

.... [2]

c) Complete the sentences below.

An atom that loses or gains an electron becomes an ....

If it loses an electron, it has an overall .... charge.   [2]

2   In the early part of the 20th century, Geiger and Marsden investigated the paths taken by positively charged alpha particles into and out of a very thin piece of gold foil. **Figure 1** shows the paths of three alpha particles.

Explain the different paths, **A**, **B** and **C**, of the alpha particles.

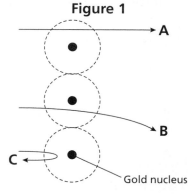

Figure 1

Gold nucleus

....

....

....

.... [3]

Total Marks .... / 10

Nuclear Radiation

1 Here is some information about potassium.

> Potassium is a metallic element in Group 1 of the Periodic Table.
> It has an atomic number of 19.
> Its most common isotope is potassium-39, $^{39}_{19}$K.
> Another isotope, potassium-40, $^{40}_{19}$K, is a radioactive isotope.

a) What is meant by 'radioactive isotope'?

_____ [1]

b) During radioactive decay, atoms of potassium-40 change into atoms of a calcium-40 by emitting an electron from the nucleus.

What type of radioactive decay has taken place?

Answer _____ [1]

c) Potassium-39 does not undergo radioactive decay.

What does this tell us about potassium-39?

_____ [1]

d) Sodium-24 is another radioactive isotope.
It decays by gamma emission.

Give the name of the element formed when this decay takes place.

Answer _____ [1]

2 Give the unit that is used to measure the activity of a radioactive isotope.

Answer _____ [1]

3 List the decay mechanisms, **alpha**, **beta** and **gamma**, in order of penetrating power.
Start with the most penetrative.

_____ [1]

Total Marks _____ / 6

Half-Life

1 Iodine is found naturally in the world and is essential to life.
It is used by the thyroid gland for the production of essential hormones.
Iodine-127 is not radioactive but iodine-131 is.
Iodine-131 has as a half-life of 8 days.

a) During the Chernobyl nuclear disaster in 1986, an explosion caused a large quantity of the isotope iodine-131 to be released into the atmosphere.
Iodine-131 from the disaster is no longer a threat to us today.

Explain why.

....

....

.... **[3]**

b) Potassium has an atomic number of 19. It decays by emitting an electron from the nucleus.

Use this information to produce a balanced decay equation for the decay of potassium-40 into caesium-40.

.... **[2]**

c) The Isotope, caesium-137, was also released during the Chernobyl disaster.
In 2046, the activity of the caesium released will be $\frac{1}{4}$ of the activity immediately after the explosion.

Calculate the half-life.

Half-life = ... **[2]**

d) Caesium-137 is a gamma emitter.

Describe gamma radiation in terms of its range in air and ionising power.

....

.... **[2]**

Total Marks .... / 9

Notes

Collins

GCSE
COMBINED SCIENCE
Biology: Paper 1 Foundation Tier

F

Materials

Time allowed: 1 hour 15 minutes

> **For this paper you must have:**
> - a ruler
> - a calculator.

Instructions

- Answer **all** questions in the spaces provided.
- Do all rough work on the page. Cross through any work you do not want to be marked.

Information

- There are **70** marks available on this paper.
- The marks for each question are shown in brackets [ ].
- You are expected to use a calculator where appropriate.
- You are reminded of the need for good English and clear presentation in your answers.
- When answering questions 03.3 and 05.4 you need to make sure that your answer:
  - is clear, logical, sensibly structured
  - fully meets the requirements of the question
  - shows that each separate point or step supports the overall answer.

Advice

- In all calculations, show clearly how you work out your answer.

01 Gonorrhoea is a disease caused by a microorganism.

Figure 1 shows the microorganism that causes gonorrhoea.

Figure 1

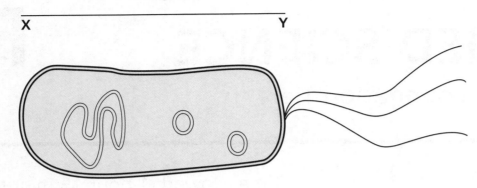

01.1 What type of microorganism is this?

Tick **one** box.

Bacterium   ☐

Fungus   ☐

Protist   ☐

Virus   ☐

[1 mark]

01.2 The magnification of **Figure 1** is ×14000.

The length of the microorganism is shown by line **XY**.

$$\text{magnification} = \frac{\text{size of image}}{\text{size of real object}}$$

What is the real length of the microorganism in millimetres?

Real length = _____ mm **[2 marks]**

02 **Figure 2** shows the number of people who tested positive for salmonella food poisoning each month in one year.

Figure 2

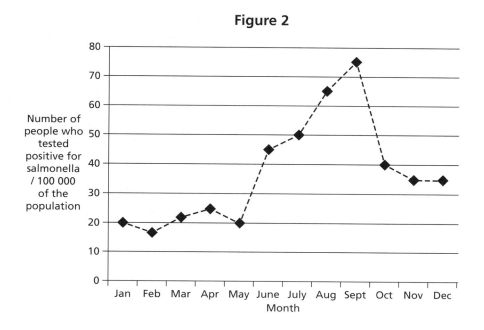

02.1 Which **two** symptoms are people who text positive for salmonella likely to show?

Tick **two** boxes.

Vomiting ☐   Abdominal cramps ☐

Red skin rash ☐   Discharge from the vagina or penis ☐   **[2 marks]**

02.2 Which **three** months contain the highest number of cases of salmonella?

_____ **[1 mark]**

02.3 Using your knowledge of how salmonella is passed on, suggest reasons why these months have the highest number of cases.

_____

_____

_____ **[2 marks]**

Turn over for the next question

03 Doctors are hoping that in the future they will be able to treat heart disease by injecting stem cells into a damaged heart.

03.1 How can stem cells help in the treatment of heart disease?

Tick **one** box.

They supply oxygen to the heart muscle cells. ☐

They destroy the microorganisms that cause heart disease. ☐

They can replace the damaged heart muscle cells. ☐

They form new heart valves. ☐ **[1 mark]**

Figure 3 shows where the stem cells are injected.

Figure 3

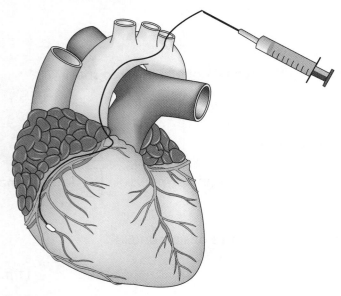

03.2 Which chamber of the heart are the cells being injected into?

Tick **one** box.

Left ventricle ☐

Right ventricle ☐

Left atrium ☐

Right atrium ☐ **[1 mark]**

03.3 In coronary heart disease, the heart muscle cells stop beating normally.

Explain what causes the cells to stop beating correctly.
Include some risk factors that may lead to this problem.

_____

_____

_____

_____

_____

_____

_____

_____

_____

[6 marks]

The stem cells for this treatment can be taken from an embryo that has been produced by cloning the patient's own cells.

03.4 Explain the benefit of using stem cells from an embryo cloned from the patient's own cells.

_____

_____

[2 marks]

03.5 Suggest why some people may object to this treatment.

_____

_____

[2 marks]

Turn over for the next question

04 Penicillin is an antibiotic drug.

04.1 Draw **one** line to link the person who discovered penicillin to the organism that it comes from.

person	organism
Mendel	fungus
Fleming	willow
Wallace	foxglove

[2 marks]

04.2 Some strains of bacteria have developed resistance to penicillin.
As more bacteria develop resistance, there is more pressure on scientists to produce new drugs.

Here are four steps carried out in the testing of new drugs.

Double-blind trials on patients	
Varying doses given to healthy volunteers	
Testing on live animals	
Low doses given to healthy volunteers	

Write the numbers **1**, **2**, **3** or **4** in the boxes to show the order of these steps in drug testing.

[3 marks]

05 Sunflower plants, such as the ones shown in **Figure 4**, can grow well in the UK.

They often grow up to three metres tall.

Figure 4

05.1 Suggest why it is an advantage for sunflowers to be taller than the other plants growing around them.

....

....

....

....

[3 marks]

05.2 Growing tall means that sunflower plants have to transport water several metres up to the leaves from the roots.

Describe the route that the water takes from the soil to the leaves.

....

....

....

....

[3 marks]

Question 5 continues on the next page

Figure 5 shows an aloe plant.

Aloe grows in areas where there is limited food for animals.

Figure 5

05.3 Use **Figure 5** to explain how the plant can survive being eaten by animals in these areas.

....

....

.... **[2 marks]**

Table 1 gives some differences between the leaves of aloe plants and sunflowers.

Table 1

Plant	Thickness of Waxy Cuticle (micrometres)	Number of Stomata (per mm²)
aloe	14.8	25
sunflower	6.1	150

05.4 Aloe grows in the desert, where it is very dry.

Use the data in **Table 1** to explain how aloe can survive in areas where it is very dry.

....

....

....

....

[4 marks]

06 Racehorses and athletes are both trained to run in races.

Figure 6

Horses can be tested to see how fit they are.

A horse's heart rate is measured when it is running at different speeds.

Some results for a horse are shown in **Table 2**.

Table 2

Speed of the Horse (Kilometres per Hour)	Heart Rate (Beats per Minute)
30	162
35	170
39	180
45	192

06.1 Plot the data from **Table 2** on the graph in **Figure 7**.

Finish the graph by drawing the line of best fit.

Figure 7

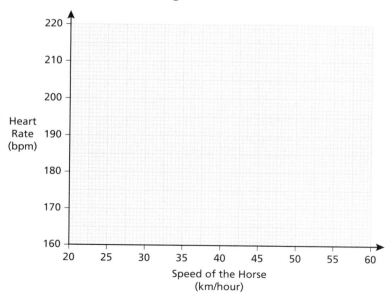

[5 marks]

Question 6 continues on the next page

06.2 Describe the pattern shown in the graph.

....

....

.... [2 marks]

06.3 Above 200 heart beats per minute, a horse starts to use **anaerobic** respiration.

Use the graph to estimate the maximum speed at which this horse can run **without** using anaerobic respiration.

Show on the graph how you work out your answer.

Maximum speed = ... km/hour [2 marks]

06.4 Write the word equation to show the reaction for **anaerobic** respiration in horses and athletes.

.... [2 marks]

06.5 Athletes and horses run better when they use **aerobic** rather than **anaerobic** respiration.

Tick **two** boxes that explain why.

Aerobic respiration releases more energy. ☐

Aerobic respiration uses less glucose. ☐

Aerobic respiration does not release carbon dioxide. ☐

Aerobic respiration does not produce lactic acid. ☐ [2 marks]

07   **Figure 8** shows six parts of the digestive system labelled **A** to **F**.

Figure 8

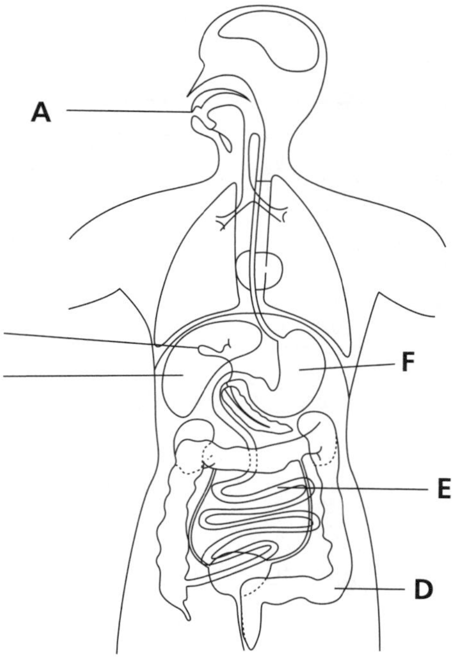

07.1 Which letter shows where bile is produced?

Answer: ....   **[1 mark]**

07.2 Which functions from the list are carried out by structure **F**?

Tick **two** boxes.

Digests food   ☐

Filters the blood   ☐

Kills microorganisms in food   ☐

Removes oxygen from the blood   ☐   **[2 marks]**

Question 7 continues on the next page

Food tests are carried out on four different food samples labelled **A**, **B**, **C** and **D**. The results are shown in **Table 3** below.

Table 3

Food sample	Results when the foods are tested		
	Starch test (iodine solution)	Glucose test (Benedict's solution)	Protein test (biuret test)
A	black	blue	blue
B	brown	blue	purple
C	brown	orange	blue
D	brown	orange	blue

07.3 Which of the three food tests involves heating the reagents?

[1 mark]

07.4 Which **two** food samples contain glucose?

[2 marks]

07.5 What colour does iodine solution turn when starch is present?

[1 mark]

07.6 Write down the name of **one** enzyme that digests protein.

[1 mark]

07.7 Write down the name of the product formed when protein is digested.

[1 mark]

08 A student wants to find out the concentration of the cell contents of potato.
They design an experiment involving osmosis.

08.1 Put a ring around the word(s) that best complete each sentence about osmosis.

Osmosis is the movement of **water / gases / sugar**.

The movement is from a **dilute / concentrated** solution to a more **dilute / concentrated** solution.

This happens through a **selectively permeable / fully permeable / impermeable** membrane.

[4 marks]

The student cuts cylinders from a potato and weighs each cylinder.
They then place each cylinder in a test tube.
Each test tube contains a different concentration of sugar solution.
After several hours the student removes the cylinders from the solutions and reweighs them.
They then calculate the percentage change in mass for each cylinder.

Figure 9 shows the student's results.

Figure 9

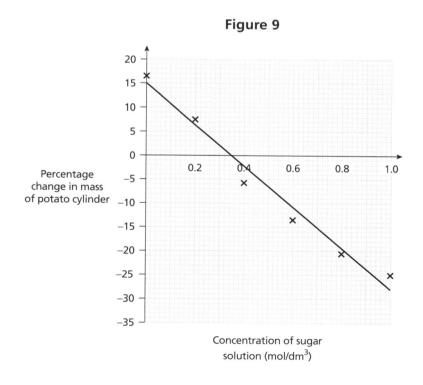

Concentration of sugar solution (mol/dm³)

Question 8 continues on the next page

08.2 The potato cylinders in 0.4 to 1.0 mol/dm³ sugar solution all lost mass. This is because water passed out of the potato tissue.

Explain why.

_____

_____

_____   **[2 marks]**

08.3 If the solution has the same concentration as the potato tissue, then there is no change in mass.

Use **Figure 9** to estimate this concentration.

Concentration = _____ mol/dm³ **[1 mark]**

08.4 **Table 4** shows possible variables in the students experiment.

Put the letter **I**, **D** or **C** in the table to show the type of variable.

I = independent variable
D = dependent variable
C = any variables that should have been controlled

Table 4

The concentration of the sugar solution	
The volume of the sugar solution	
The change in mass of the potato cylinder	
The time that each cylinder was left to soak	

[4 marks]

END OF QUESTIONS

Collins

GCSE
COMBINED SCIENCE   F
Biology: Paper 2 Foundation Tier

Materials

Time allowed: 1 hour 15 minutes

For this paper you must have:

- a ruler
- a calculator.

Instructions

- Answer **all** questions in the spaces provided.
- Do all rough work on the page. Cross through any work you do not want to be marked.

Information

- There are **70** marks available on this paper.
- The marks for each question are shown in brackets [ ].
- You are expected to use a calculator where appropriate.
- You are reminded of the need for good English and clear presentation in your answers.
- When answering question 08.3 you need to make sure that your answer:
  - is clear, logical, sensibly structured
  - fully meets the requirements of the question
  - shows that each separate point or step supports the overall answer.

Advice

- In all calculations, show clearly how you work out your answer.

Biology Practice Exam Paper 2

01 This question is about skin sensitivity.

Tim presses Kate's skin with two pin heads, as shown in **Figure 1**.

Sometimes Tim presses both pin heads on to Kate's skin and sometimes just one.
Every time he presses her skin, Kate says the number of points she feels.
Tim writes down how many times she calls the correct number of points.
He does this 20 times.

Tim does this experiment with the pins different distances apart and on three parts of the body.
The graph in **Figure 2** shows his results.

Figure 1

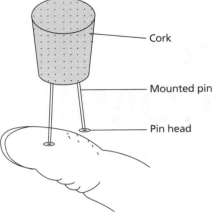

Cork

Mounted pin

Pin head

Figure 2

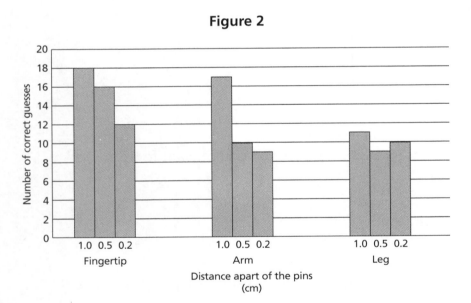

01.1 Name the part of Kate's body where her skin was most sensitive to the stimulus.

[1 mark]

01.2 Explain why Tim touched Kate's skin 20 times in each trial.

[2 marks]

01.3 What do the results from Kate's arm show?

....

.... **[2 marks]**

01.4 Kate's response to the pins was not a reflex action.

Explain how you can tell this.

....

.... **[2 marks]**

01.5 Figure 3 shows a reflex arc.

Figure 3

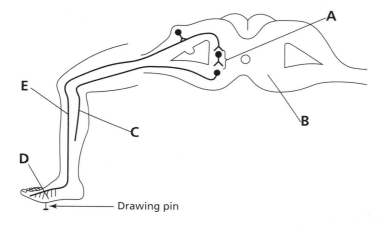

Drawing pin

Complete **Table 1** by matching a letters on the diagram to each part of the reflex arc.
The first one has been done for you.

Table 1

Part of Reflex Arc	Letter
relay neurone	A
receptor	
sensory neurone	
spinal cord	
motor neurone	

[3 marks]

Turn over for the next question

02 Polar bears (scientific name *Ursus maritimus*) live in the Arctic.
They need to move about on the ice and they only eat seals.

Figure 4

02.1 Write down **two** features shown in the **Figure 4** that make polar bears well
adapted for hunting on ice and eating seals.

....

.... **[2 marks]**

Alaskan bears (*Ursus arctos*) live in Alaska, south of the Arctic.
They catch fish for food.

Figure 5

02.2 What genus do the Alaskan bear and polar bear belong to?

.... **[1 mark]**

02.3 The temperature in Alaska is getting warmer.
Alaskan bears are now living in some of the same areas as polar bears.

Which of these statements are true?

Tick **two** boxes.

Polar bears and Alaskan bears will compete for food. ☐

Polar bears will mate with Alaskan bears and the offspring will be a new species. ☐

Polar bears and Alaskan bears may mate and produce sterile hybrids. ☐

Polar bears and Alaskan bears cannot mate because they are in different kingdoms. ☐

The habitats of the polar bears and the Alaskan bears may overlap. ☐

[2 marks]

The level of carbon dioxide in the air is increasing.
Scientists think that this could be due to human activity.

02.4 Suggest how human activity could cause the level of carbon dioxide in the atmosphere to increase.

_____

_____

_____

_____

[3 marks]

02.5 Scientists think that increasing levels of carbon dioxide could cause the temperature in the Arctic to increase.

Explain how this could happen.

_____

_____

_____

_____

[3 marks]

Turn over for the next question

03 Copper is an important element in many organisms. However, in high concentrations it is poisonous.

Figure 6 shows waste soil from a copper mine piled up next to a river.

Figure 6

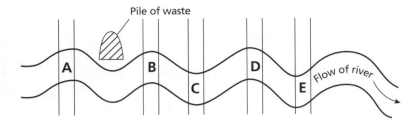

A scientist investigated copper concentrations in the river.
They took samples of water from locations **A**, **B**, **C**, **D** and **E**.
They then measured the concentration of copper in each sample.

The results are shown in **Table 2**.

Table 2

Sample Location	Concentration of Copper (Micrograms per Litre)
A	1
B	16
C	10
D	7
E	3

03.1 Which sample, **A**, **B**, **C**, **D** or **E**, contained the highest concentration of copper?
Explain why.

....

.... [2 marks]

03.2 The scientist noticed that there were fewer plants growing on the river bank at
location **B** than at location **A**.

Suggest an explanation for this difference.

....

.... [2 marks]

The scientist takes sample plants from each location and continues to grow them in the laboratory.

The scientist waters the plants regularly with water containing copper.

The survival rate of the plants is shown in **Table 3**.

Table 3

Location That Plants Were Taken From	Survival (%)
A	1
B	100
C	90
D	55
E	20

03.3 If the scientists grew 200 plants from site **A**, how many would have survived?

Tick **one** box.

1 ☐

2 ☐

100 ☐

200 ☐

[1 mark]

03.4 Why is there such a large difference in the survival rate of plants from location **A** compared to plants from location **B**?

....

....

[2 marks]

Question 3 continues on the next page

03.5 The scientist wrote an explanation for why 100% of the plants from B could survive the watering.

Complete the scientist's explanation by choosing words from this list.

mutation	resistant	reproduce	natural selection
died	sensitive	denaturing	seed

A plant growing at B must have had a _____ that allowed it

to survive.

It could therefore _____ and pass on its genetic material.

After many generations, all the plants growing at B were _____

to copper.

This process is called _____ .   **[4 marks]**

04 In 1866, Gregor Mendel published a paper on genetics.

Figure 7

Mendel carried out breeding experiments with pea plants.
He crossed tall pea plants with dwarf pea plants.
He collected the seeds and planted them.
He then self-pollinated the plants that grew and again planted the seeds that were made.

His results are shown in **Figure 8**.

Figure 8

Parents	TT Tall		tt Dwarf
Gametes	T		t
First generation		Tt All Tall	
Gametes	T or t		T or t
Second generation	TT Tall	Tt Tall	Tt Tall	tt Dwarf

04.1 Complete the sentences about Mendel's results.

Use words from the box.

dominant dwarf genotypes heterozygous homozygous phenotypes recessive tall

The genotype of all the plants in the first generation is ....

The phenotype of these plants is ....

This is because the tall allele is .... over the dwarf allele.

The second generation contains plants that have three

different .... **[4 marks]**

Question 4 continues on the next page

04.2 To test his ideas, Mendel crossed a plant from the first generation with a dwarf plant.

The genetic diagram in **Figure 9** shows this cross.

Figure 9

	Tt	
	T	**t**
tt	**t**		

Complete the genetic diagram.
Use the completed diagram to write down the ratio of phenotypes that this cross would produce.

Ratio = _____ **[4 marks]**

05 Here is an advert of a DNA testing kit.

> **BABY TESTING KIT**
>
> Do you want to know the sex of your baby before it is born?
> This testing kit can answer that question.
> When a woman is pregnant, a very small amount of the baby's DNA is in the mother's blood.
> This means a small drop of the mother's blood can be tested to see if it has any DNA from a Y chromosome.
> This will tell you if you are going to have a boy or a girl.

05.1 Choose words from the advert to complete these definitions.

The chemical that contains the genetic code is called ....

A thin strand of genetic material is called a ....

A length of genetic material that codes for a protein is

called a ....   **[3 marks]**

05.2 The test involves finding out if any DNA in the mother's blood is from a Y chromosome.

Explain how this can be used to tell the sex of the baby.

....

....   **[2 marks]**

05.3 At present, scientists do this test on cells taken directly from the baby inside the mother's uterus.
They hope that in the future they will be able to carry out the same tests using samples of the mother's blood.

Suggest **one** reason why scientists think that this would be a better method.

....

....   **[1 mark]**

Turn over for the next question

06 *Acetabularia* is a single-celled organism that lives in the sea.

It has an unusual shape, as shown in **Figure 10**.

Figure 10

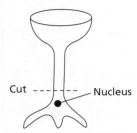

Cut - - - - - Nucleus

06.1 When *Acetabularia* reproduces, the nucleus and then the cytoplasm divides into two.

What is this type of reproduction called?

_____ **[1 mark]**

06.2 *Acetabularia* produced by this type of reproduction always have the same shaped cap as the parent.

Why is this?

_____ **[1 mark]**

06.3 A scientist called Hammerling performed some experiments on *Acetabularia*.
He cut the organism into two as shown on **Figure 10**.
He found that the bottom part of the organism survived and grew a new top (cap).

Explain why the bottom part of the organism could grow a new cap.

_____

_____

_____ **[2 marks]**

07 A woman draws a circle on her calendar when she starts her period (menstruation).

Figure 11

March
S M T W T F S
(1) 2 3 4 5
6 7 8 9 10 11 12
13 14 15 16 17 18 19
20 21 22 23 24 25 26
27 28 29 30 31

07.1 What happens to a woman's uterus during a period?

....

.... **[2 marks]**

07.2 The menstrual cycle of this woman normally lasts 28 days.

Mark on the calendar when the woman would expect her next period to start.   **[1 mark]**

The woman does not want to get pregnant.
She thinks that avoiding intercourse between 14th to 16th March will stop her getting pregnant.

07.3 Suggest why she might think this.

....

.... **[2 marks]**

07.4 Why is avoiding intercourse on these days not a reliable method of contraception?

.... **[1 mark]**

Question 7 continues on the next page

07.5 Write down **two** more reliable methods of contraception that the woman could use.

....

.... **[2 marks]**

The woman's menstrual cycle is controlled by hormones.

07.6 Name **two** hormones that control the menstrual cycle.

.... **[2 marks]**

07.7 Where in the female body are these hormones produced?

.... **[1 mark]**

08   This question is about hormones and homeostasis.

Homeostasis means keeping the internal conditions of the human body constant.

08.1 The blood glucose level needs to be kept constant in the body.

Write down **two** other factors that need to be controlled in the body.

....

.... [2 marks]

08.2 Hormones are important in homeostasis.

What is a hormone?

....

.... [2 marks]

Question 8 continues on the next page

The article in **Figure 12** appeared in a recent newspaper.

Figure 12

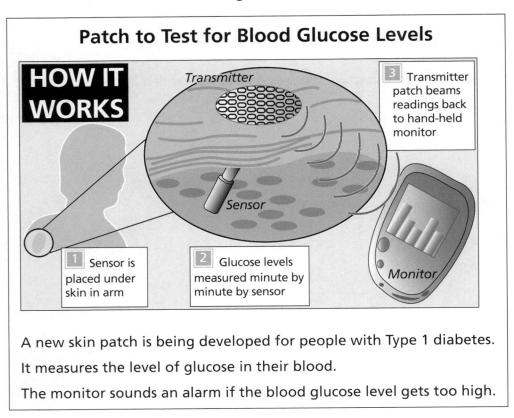

Patch to Test for Blood Glucose Levels

HOW IT WORKS

Transmitter

3 Transmitter patch beams readings back to hand-held monitor

Sensor

Monitor

1 Sensor is placed under skin in arm

2 Glucose levels measured minute by minute by sensor

A new skin patch is being developed for people with Type 1 diabetes.

It measures the level of glucose in their blood.

The monitor sounds an alarm if the blood glucose level gets too high.

08.3 If the monitor sounds an alarm, what does the person with Type 1 diabetes need to do?

Explain your answer.

....

....

....

....

[5 marks]

END OF QUESTIONS

Collins

GCSE
COMBINED SCIENCE
Chemistry: Paper 1 Foundation Tier

F

Materials

Time allowed: 1 hour 15 minutes

For this paper you must have:

- a ruler
- a calculator
- the periodic table (see page 168).

Instructions

- Answer **all** questions in the spaces provided.
- Do all rough work on the page. Cross through any work you do not want to be marked.

Information

- There are **70** marks available on this paper.
- The marks for each question are shown in brackets.
- You are expected to use a calculator where appropriate.
- You are reminded of the need for good English and clear presentation in your answers.
- When answering question 05.2 you need to make sure that your answer:
  - is clear, logical, sensibly structured
  - fully meets the requirements of the question
  - shows that each separate point or step supports the overall answer.

Advice

- In all calculations, show clearly how you work out your answer.

Chemistry Practice Exam Paper 1

01 This question is about the atomic model.

Figure 1 shows a model of a helium atom.

Figure 1

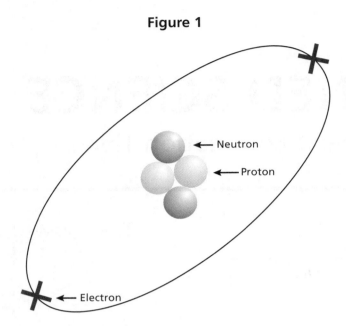

01.1 What are the relative electric charges on the particles in an atom?

Tick **one** box.

Proton	Neutron	Electron	
0	−1	+1	☐
+1	0	−1	☐
−1	0	0	☐
+1	+1	−1	☐

[1 mark]

01.2 A helium atom has an overall neutral charge.

State why.

_____

_____

[1 mark]

01.3 What is the mass number and atomic number of helium?

Tick **one** box.

Mass Number	Atomic Number	
2	2	☐
6	2	☐
6	4	☐
4	2	☐ [1 mark]

01.4 Why is helium an unreactive element?

Tick **one** box.

It has an equal number of protons and neutrons. ☐

Elements with two electrons in their outer shell are unreactive. ☐

It is a gas at room temperature and pressure. ☐

It has a full outer shell of electrons. ☐ [1 mark]

01.5 An isotope of helium has only one neutron in its atoms.

Which statements about how an atom of this isotope compares to the atom in **Figure 1** are true?

Tick **two** boxes.

It has the same atomic number. ☐

It has a higher mass. ☐

It has a different atomic number. ☐

It has a different mass number. ☐

It has the same mass number. ☐ [2 marks]

Turn over for the next question

02 Iron is found in the Earth as the compound iron oxide (Fe_2O_3).

02.1 How many atoms are in one molecule of Fe_2O_3?

Answer: _____ **[1 mark]**

Iron is extracted from iron(III) oxide using carbon in the following reaction:

iron(III) oxide + carbon → iron + carbon dioxide

$$2Fe_2O_3 \quad + \quad 3C \quad \rightarrow 4Fe + \quad 3CO_2$$

02.2 Calculate the relative formula mass (M_r) of carbon dioxide (CO_2).

Relative atomic masses (A_r): carbon = 12; oxygen = 16

Answer: _____ **[1 mark]**

02.3 Both oxidation and reduction take place when iron is extracted from iron(III) oxide.

Draw **one** line from each type of change to the name of the substance that is changed in that way.

Change	Substance
	iron
oxidation	iron(III) oxide
	carbon
reduction	carbon dioxide

[2 marks]

02.4 Gold does not have to undergo an extraction process.

Explain why.

....

....

....

.... **[2 marks]**

Turn over for the next question

03   **Table 1** shows some of the properties of elements in Group 7 of the periodic table.

Table 1

Element	Density (g/cm³)	Melting point (°C)	Boiling point (°C)
Fluorine	0.0017	−219.6	−188.1
Chlorine	0.0032	−101.5	−34.0
Bromine	3.1028	−7.3	58.8
Iodine	4.9330	113.7	184.3

03.1 State **one** trend in the properties of the Group 7 elements as shown in **Table 1**.

....

.... [1 mark]

03.2 Astatine is found in Group 7, below iodine.

Use the information in **Table 1** to estimate the melting point of astatine.

Answer: .... °C   [1 mark]

Group 7 elements all exist as molecules containing two atoms.

03.3 Chlorine and fluorine are both gases at room temperature.
Changing the temperature of a mixture of the gases can be used to separate them.

Complete the sentences to explain how this works.
Use words from the box.

chlorine	fluorine	liquid
solid	heat	cool

_____ the mixture to −34°C.

_____ will become a _____ and

can be removed.   **[3 marks]**

03.4 What is the formula for a molecule of bromine?

Answer: _____ **[1 mark]**

03.5 Which statement correctly describes what happens when a Group 7
element boils?

Tick **one** box.

Covalent bonds form   ☐

Intermolecular forces form   ☐

Covalent bonds break   ☐

Intermolecular forces break   ☐   **[1 mark]**

Question 3 continues on the next page

03.6 The Group 7 elements all react with Group 1 elements to form ionic compounds.

Explain why all the Group 7 elements share this chemical property.

_____

_____ **[1 mark]**

03.7 The Group 1 element sodium reacts with chlorine to form sodium chloride.
Figure 2 shows the outer electrons in an atom of sodium and in an atom of chlorine.

Figure 2

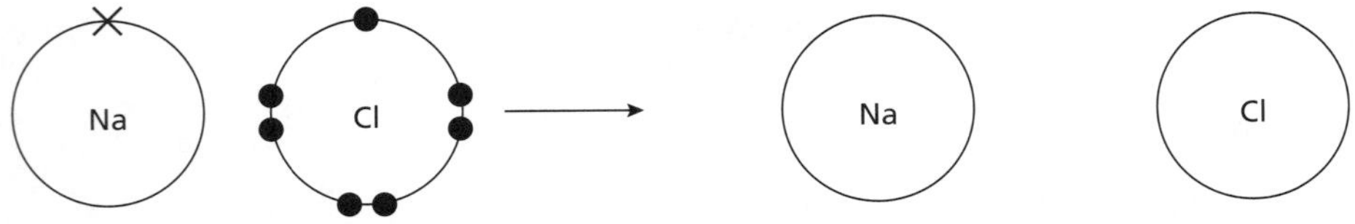

Complete the diagram to show the arrangement of electrons in sodium chloride.
You should give the formula of each ion formed. **[5 marks]**

03.8 The Group 7 elements have trends in their properties.

Which of the following correctly describe how properties change as you move
down Group 7?

Tick **two** boxes.

Boiling point increases ☐

Reactivity increases ☐

Molecular mass increases ☐

Melting point decreases ☐ **[2 marks]**

Workbook

04 A student was asked to produce copper(II) sulfate crystals.
They used the method shown in **Figure 3**.

Figure 3

Step 1 labels: Spatula, (black) Copper(II) oxide, Glass rod to stir, Copper(II) oxide being stirred to react with sulfuric acid

Step 2 labels: Filter paper, Residue of copper(II) oxide left behind, Conical flask, Blue copper(II) sulfate solution

Step 3 labels: Copper(II) sulfate crystals, Evaporating dish

Step 1   **Step 2**   **Step 3**

04.1 Give the name of the separation process used in **Step 2** and explain
why it was used.

....

....

....

....

[2 marks]

04.2 Identify one hazard in **Step 1** or **Step 2** and suggest a method of
reducing the risk.

Hazard: ....

Way of reducing the risk: ....

[2 marks]

Question 4 continues on the next page

04.3 Another student was asked to produce the salt calcium chloride.

Name the metal oxide and acid they should use.

Metal oxide: .... **[1 mark]**

Acid: .... **[1 mark]**

04.4 The ions in magnesium chloride are Ca^{2+} and Cl^-.

State the formula of calcium chloride.

.... **[1 mark]**

05   **Figure 4** shows the structure of graphite.

Figure 4

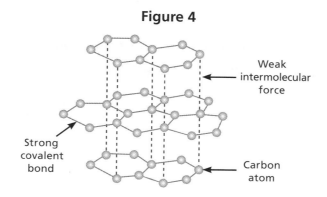

05.1 Graphite and diamond are both giant covalent compounds.

State **one difference** and **one similarity** between the structures of diamond and graphite.

Similarity: ....

Difference: .... **[2 marks]**

05.2 Graphene is a single layer of graphite.
Its properties include being an electrical conductor, strong and transparent.
In the future, it could be used to make touch-screens for electronic devices like mobile phones.

Explain why graphene has these properties in terms of its structure and why it is a good choice of material for a touch-screen.

....

....

....

....

....

....

....

[6 marks]

Turn over for the next question

06 A student was asked to carry out the electrolysis of copper(II) sulfate solution using inert graphite electrodes.

06.1 Complete **Figure 5** to show the apparatus they should use.
Label the following on your completed diagram:
- Anode
- Cathode.

Figure 5

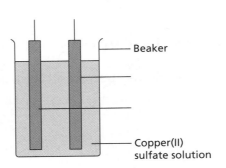

Beaker

Copper(II)
sulfate solution

[3 marks]

06.2 Explain why copper(II) sulfate must be dissolved in water in order for electrolysis to take place.

....

....

.... [2 marks]

06.3 Hydrogen (H^+) ions and copper (Cu^{2+}) ions are both attracted to the cathode but only one product is formed.

Predict the name of the product formed at the cathode.
Explain why only this product is formed.

....

.... [2 marks]

07   Hydrochloric acid neutralises sodium hydroxide solution.

The equation for the reaction is:

$HCl(aq) + NaOH(aq) \rightarrow NaCl(aq) + H_2O(l)$

07.1   The pH scale is a measure of the acidity or alkalinity of a solution.

Draw lines from each reactant to its pH value.

hydrochloric acid		1
		4
		7
sodium hydroxide		13

[2 marks]

07.2   Name the **two** products of the reaction.

.... [2 marks]

07.3   A student was asked to carry out this reaction.

Describe the method they could use to tell when all the sodium hydroxide had reacted with hydrochloric acid.

....

....

....

....

[3 marks]

Question 7 continues on the next page

Another student repeated the reaction but added the hydrochloric acid to the sodium hydroxide 2.5 cm³ at a time.

They measured the pH of the mixture each time using a pH probe.

Figure 6 shows the graph of their results.

Figure 6

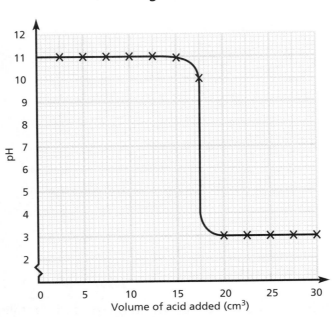

07.4 State the volume of sodium hydroxide needed to neutralise the acid.

Volume = _____ cm² **[1 mark]**

07.5 What else could the student do to be more certain of the volume needed?

_____

_____

_____ **[2 marks]**

08 A student investigated how the mass of magnesium changes when it reacts with dilute hydrochloric acid.

Figure 7 shows the apparatus they used.

Figure 7

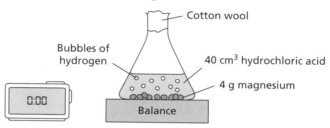

08.1 Use the correct state symbols from the box to complete the chemical equation.

| aq | g | l | s |

magnesium + hydrochloric acid → magnesium chloride + hydrogen

$$Mg \underline{\quad} + 2HCl(aq) \rightarrow MgCl_2(aq) + H_2 \underline{\quad}$$

[2 marks]

08.2 State **one** function of the cotton wool.

....

.... [1 mark]

Table 2 shows the student's results.

Table 2

Time in s	Total loss of mass in g
0	0.00
10	1.25
20	2.30
30	3.00
40	3.45
50	3.70
60	3.87
70	4.00
80	4.00
90	4.00

Question 8 continues on the next page

08.3 On **Figure 8**:
- Plot the results from **Table 2**.
- Draw a line of best fit.

Figure 8

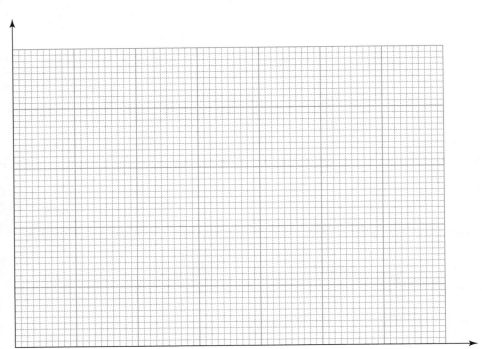

Total loss of
mass in g

Time in s

[4 marks]

08.4 At what time did the mass loss stop changing?

Answer: _____ s [1 mark]

08.5 The student observed that there was still magnesium in the flask at this time.

Explain why the mass loss stopped changing.

_____

_____ [1 mark]

08.6 Explain, in terms of particles, why the mass changes during the reaction.

_____

_____

_____

_____ [2 marks]

END OF QUESTIONS

Chemistry Practice Exam Paper 1

There are no questions printed on this page

Collins

GCSE
COMBINED SCIENCE
Chemistry: Paper 2 Foundation Tier

F

Materials

Time allowed: 1 hour 15 minutes

For this paper you must have:

- a ruler
- a calculator
- the periodic table (see page 168).

Instructions

- Answer **all** questions in the spaces provided.
- Do all rough work on the page. Cross through any work you do not want to be marked.

Information

- There are **70** marks available on this paper.
- The marks for each question are shown in brackets.
- You are expected to use a calculator where appropriate.
- You are reminded of the need for good English and clear presentation in your answers.
- When answering questions 04 and 05.2 you need to make sure that your answer:
  - is clear, logical, sensibly structured
  - fully meets the requirements of the question
  - shows that each separate point or step supports the overall answer.

Advice

- In all calculations, show clearly how you work out your answer.

01 This question is about atmospheric pollutants from fuels.

Methane (CH_4) is a hydrocarbon which is used as a fuel.

The equation shows the reaction for the complete combustion of methane:

methane + oxygen → carbon dioxide + water

$$CH_4 + 2O_2 \rightarrow CO_2 + XH_2O$$

01.1 X represents what number?

Tick **one** box.

1 ☐

2 ☐

3 ☐

4 ☐

[1 mark]

01.2 Tests can be used to identify gases.

Draw **one** line from each gas to the test that can be used to identify it.

| | Hold a burning splint at the open end of a test tube containing the gas. |

carbon dioxide

| Hold a glowing splint at the open end of a test tube containing the gas. |

| Bubble the gas through limewater. |

oxygen

| Put damp litmus paper into the gas. |

[2 marks]

01.3 If there is a limited amount of oxygen, then methane will undergo incomplete combustion. One product is carbon monoxide.

Which other product may be produced in this reaction?

Tick **one** box.

Carbon particles ☐

Sulfur dioxide ☐

Nitrogen oxide ☐

Hydrogen ☐

[1 mark]

01.4 Give **two** reasons why it is difficult to detect the toxic gas carbon monoxide.

_____

_____ [2 marks]

01.5 Increased amounts of pollutants in the air cause problems.

Draw **one** line from each pollutant to the problem it causes.

Pollutant	Problem
	global warming
sulfur dioxide	global dimming
	acid rain
carbon particles	destruction of ozone layer

[2 marks]

Turn over for the next question

Chemistry Practice Exam Paper 2

02   Humans need clean drinking water.
Water that is safe to drink is called potable water.

02.1 What is the name of the process that produces potable water from seawater?

Tick **one** box.

Anaerobic digestion ☐

Decomposition ☐

Desalination ☐

Neutralisation ☐   [1 mark]

02.2 Potable water **cannot** be described as pure.

Explain why.

_____

_____

_____   [2 marks]

02.3 Name **one** sterilising agent used to kill microorganisms in water.

_____   [1 mark]

02.4 Fluoride is sometimes added to water.

A sample of drinking water contains 1.35 mg of fluoride per dm^3 of water.

Calculate the mass of fluoride in 250 cm^3 of water.
Give your answer to 2 decimal places.
1000 cm^3 = 1 dm^3

Mass = _____ mg   **[3 marks]**

Turn over for the next question

03 Cracking can be carried out in the laboratory using the apparatus shown in **Figure 1**.

Figure 1

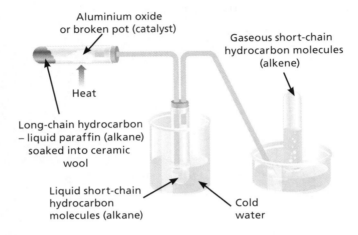

One example of a cracking reaction is:

$C_{10}H_{22} \rightarrow C_7H_{16} + M$

03.1 State the formula of **M**.

Formula: _____ **[1 mark]**

03.2 Describe a test to show that the gas produced is an alkene.

Give the expected result of the test.

_____

_____

_____ **[2 marks]**

An alkene called ethene is used to make a polymer called poly(ethene).
There are two types of poly(ethene): low density (LD) and high density (HD).

Table 1 shows the properties of the two types of poly(ethene).

Table 1

Property	LD Poly(ethene)	HD Poly(ethene)
Density	0.91–0.94 g/cm³	0.95–0.97 g/cm³
Flexibility	High	Low
Strength	Low	High

03.3 A manufacturer wants to make plastic buckets.

Suggest what type of poly(ethene) they should use.
Give a reason for your choice.

....

....

.... **[2 marks]**

03.4 **Figure 2** shows the structure of a polymer.

Figure 2

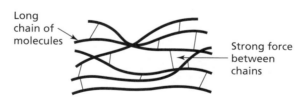

Long chain of molecules

Strong force between chains

Explain why polymers are solids at room temperature.

....

....

.... **[3 marks]**

Turn over for the next question

04   An increase in average global temperature is a cause of climate change.

Explain the effects of global climate change on the environment, humans and wildlife.

_____

_____

_____

_____

_____

_____

_____

_____

_____

_____

_____

_____

_____

                                                         [6 marks]

05 Steel and aluminium are used to make drinks cans.
They are both limited resources.

05.1 Explain what is meant by 'limited resources'.

....

.... **[1 mark]**

05.2 The metal from drinks cans can be recycled.

The following steps are used:
1. The cans are collected and sorted.
2. They are melted down and then cooled to form blocks of metal.
3. The blocks are rolled into thin sheets, which can be used to make new products.

Evaluate the use of recycling metal cans as a method of reducing the use of limited resources.

....

....

....

....

....

....

.... **[6 marks]**

Turn over for the next question

06 Magnesium reacts with dilute hydrochloric acid:

magnesium + hydrochloric acid → magnesium chloride + hydrogen

$$Mg(s) \quad + \quad 2HCl(aq) \quad \rightarrow \quad MgCl_2(aq) \quad + \quad H_2(g)$$

A student investigated how the volume of hydrogen produced over time changes when magnesium is reacted with two different concentrations of dilute hydrochloric acid.

This is the method used:
1. Measure 20 cm³ of 0.5 mol/dm³ hydrochloric acid using a measuring cylinder.
2. Pour the acid into the conical flask.
3. Add 2 g of magnesium strip.
4. Place the bung into the flask.
5. Measure the volume of gas every 30 seconds until the reaction is complete.
6. Repeat using 1 mol/dm³ hydrochloric acid.

The student reached the end of **Step 2** and set up the apparatus as shown in **Figure 3**.

Figure 3

06.1 Identify what the student should do before continuing with the method.
Describe what could happen if the student continued without making any changes.
Explain how this would affect the results.

....

....

....

....

[3 marks]

The student corrected the error.

Their results are shown in **Table 2**.

Table 2

| Time in s | Total volume of hydrogen in cm³ | |
	0.5 mol/dm³ acid	1 mol/dm³ acid
0	0.0	0.0
30	8.2	14.1
60	14.4	25.5
90	20.0	33.6
120	25.1	36.8
150	29.3	37.6
180	33.7	38.0
210	36.2	38.0
240	37.8	38.0
270	38.0	38.0

06.2 On **Figure 4**:

- Plot both sets of results on the grid.
- Draw two lines of best fit.

Figure 4

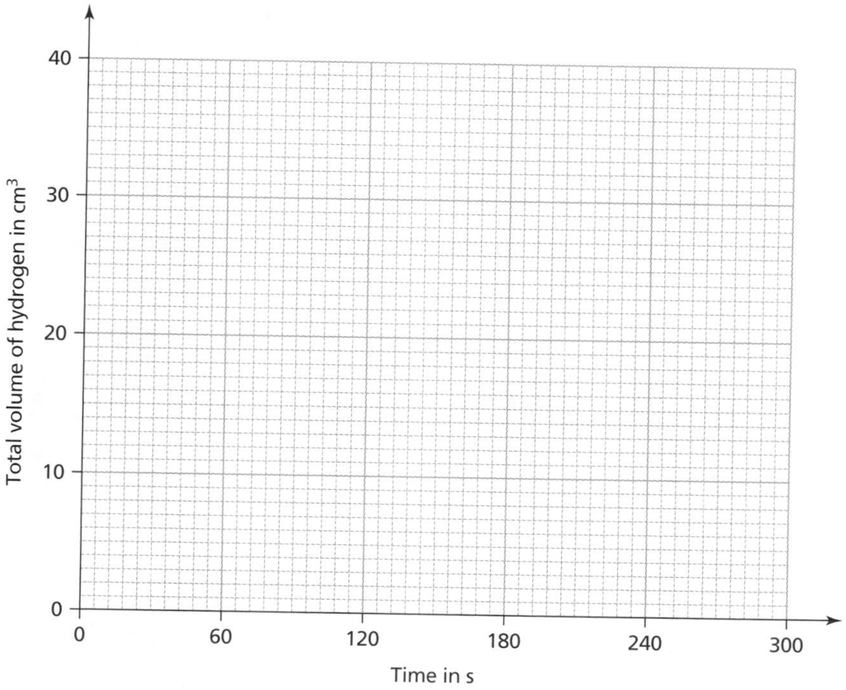

[4 marks]

Question 6 continues on the next page

06.3 How does the concentration of acid affect the rate of the reaction?

....

.... **[1 mark]**

06.4 **Figure 5** represents acid at a concentration of 0.5 mol/dm^3.
The grey dots are particles of acid.

Figure 5

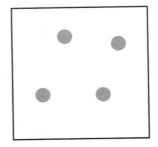

Which diagram represents acid at 1 mol/dm^3?

Tick **one** box.

A

☐

C

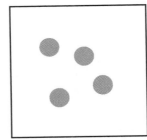

☐

B

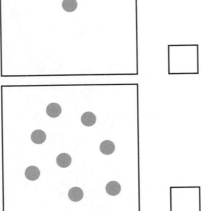

☐

D

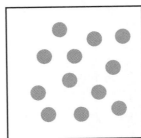

☐

[1 mark]

06.5 Explain why, in terms of particles and collisions, the concentration of acid affects the rate of the reaction.

....

....

....

....

....

.... **[3 marks]**

The student decided to research how the temperature of the acid affected the rate of reaction.

Figure 6 shows a graph the student found in a text book.

Figure 6

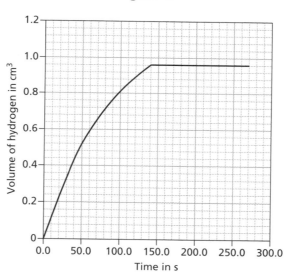

06.6 State the time at which the reaction finishes.

Answer: .... s   **[1 mark]**

Question 6 continues on the next page

06.7 The rate of this reaction can be found by measuring the volume of hydrogen produced over time.

Calculate the mean rate of the reaction.

Use the formula:

$$\text{mean rate of reaction} = \frac{\text{quantity of product formed}}{\text{time taken}}$$

Give your answer to 2 significant figures.

Rate of reaction = _____ cm³/s **[3 marks]**

07 This question is about how the amounts of the different gases in the atmosphere have changed. **Figure 7** shows how the levels of oxygen in the atmosphere have changed since the Earth was formed.

Figure 7

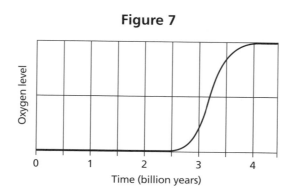

07.1 Use the graph to state when oxygen first started to be produced.

Answer: _____ billion years   **[1 mark]**

07.2 The air today is approximately one-fifth oxygen.

Calculate the approximate volume of oxygen in 200 cm³ of air.

Answer: _____ cm³   **[1 mark]**

07.3 Complete the sentences to explain how the amount of oxygen in the air increased to the amount found today.
Use words from the box.

| animals | nitrogen | photosynthesis | respiration | water | algae |

Billions of years ago, _____ in the water evolved and starting

converting carbon dioxide and _____ into oxygen by the process of

_____ .

Land plants then evolved to increase amounts of oxygen in the atmosphere further.   **[3 marks]**

Question 7 continues on the next page

Chemistry Practice Exam Paper 2

Another gas found in the air today is carbon dioxide.

07.4 Describe how carbon dioxide helps to maintain temperatures on Earth.

....

....

....

....

....

[3 marks]

07.5 In what way has the amount of carbon dioxide in the atmosphere changed over the last 100 years?

....

[1 mark]

07.6 Describe **one** way in which human activity has brought about this change to the amount of carbon dioxide in the atmosphere today.

....

....

[2 marks]

08   Paper chromatography can also be used to identify substances.

Figure 8 shows the results from chromatography carried out on a mixture.

Figure 8

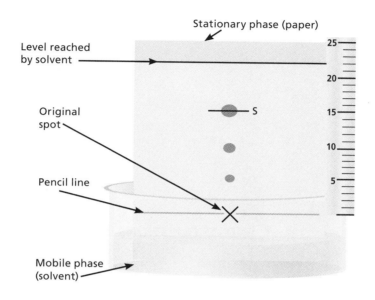

08.1 How many substances are present in the mixture?

Answer: _____ **[1 mark]**

08.2 To identify a substance its R_f value must be calculated.

This is done using the equation: $R_f = \dfrac{\textbf{distance moved by sample from the start line}}{\textbf{distance moved by solvent from the start line}}$

Calculate the R_f value of **S**.
Give your answer to 2 decimal places.

Answer: _____ **[3 marks]**

08.3 Explain how paper chromatography separates the substances in a mixture.

....

....

....

[3 marks]

END OF QUESTIONS

Collins

GCSE
COMBINED SCIENCE
Physics: Paper 1 Foundation Tier

F

Materials

Time allowed: 1 hour 15 minutes

For this paper you must have:

- a ruler
- a calculator
- the Physics Equation Sheet (page 167).

Instructions

- Answer **all** questions in the spaces provided.
- Do all rough work on the page. Cross through any work you do not want to be marked.

Information

- There are **70** marks available on this paper.
- The marks for each question are shown in brackets [ ].
- You are expected to use a calculator where appropriate.
- You are reminded of the need for good English and clear presentation in your answers.
- When answering questions 01.2 and 04.5 you need to make sure that your answer:
  - is clear, logical, sensibly structured
  - fully meets the requirements of the question
  - shows that each separate point or step supports the overall answer.

Advice

- In all calculations, show clearly how you work out your answer.

01   **Figure 1** shows what happens to each 100 joules of energy from coal that is burned in a power station.

Figure 1

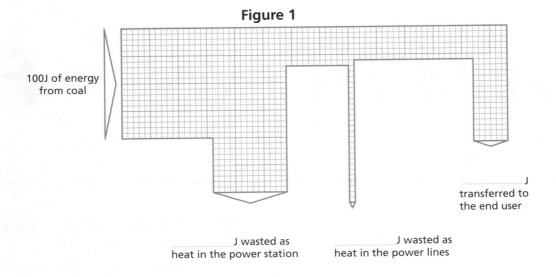

100J of energy from coal

.... J transferred to the end user

.... J wasted as heat in the power station

.... J wasted as heat in the power lines

01.1   Add the missing values to the diagram.   **[3 marks]**

01.2   For the same cost, the electricity company could:
- install new power lines that only waste half as much energy as the old ones
  OR
- use a quarter of the heat wasted at the power station to heat schools in a nearby town.

Which of these two things do you think they should do?
Give a reason for your answer.

....

....

....

....

....

[4 marks]

01.3   Calculate the efficiency of the coal powered station in **Figure 1**.

Efficiency = ... %   **[1 mark]**

02 A gas burner is used to heat some water in a pan.

By the time the water starts to boil:

- 60% of the energy released has been transferred to the water
- 20% of the energy released has been transferred to the surrounding air
- 13% of the energy released has been transferred to the pan
- 7% of the energy released has been transferred to the gas burner itself.

02.1 Some of the energy released by the burning gas is wasted.

What happens to this wasted energy?

....

.... **[2 marks]**

02.2 What percentage of the energy from the gas is wasted?

Percentage = ... % **[1 mark]**

02.3 How efficient is the gas burner at heating water?

Efficiency = ... % **[1 mark]**

Turn over for the next question

03 A book weighs 6 newtons.
A librarian picks up the book from the ground and puts it on a shelf that is 2 metres high.

03.1 Calculate the work done on the book.

Work done = _____ J   **[2 marks]**

03.2 The next person to take the book from the shelf accidentally drops it.
The book falls 2 m to the ground.

Calculate how much gravitational energy it loses as it falls.
The gravitational field strength is 10 N/kg.

Answer = _____ J   **[2 marks]**

03.3 Complete the sentence below to describe the energy transfers that take place.

As the book falls _____ energy is converted into

_____ energy.

When the book hits the floor some of this energy is then heard as

_____ energy.   **[3 marks]**

04 Electricity can be produced from a number of different energy resources.

04.1 Complete **Table 1** to show the energy transfers that occur for different resources.

Table 1

Device	Energy resource	Useful energy transfer from resource	
Coal-fired power station	Coal	chemical	⟶ electrical
Hydroelectric power station	Stored water		⟶ electrical
Solar panel	Sun		⟶ electrical
Wind turbine	Wind		⟶ electrical
Gas-fired power station	Gas		⟶ electrical

[4 marks]

04.2 State which of the five energy resources in **Table 1** are **not** renewable.

.... **[1 mark]**

04.3 Give **one** other non-renewable energy resource.

.... **[1 mark]**

04.4 Give **one** renewable energy source **not** listed in **Table 1**.

.... **[1 mark]**

Question 4 continues on the next page

04.5 State and explain the advantages and disadvantages of using nuclear power stations to produce electricity.

[4 marks]

05 A student carries out an experiment to investigate the current through component **X**.
A circuit is set up as shown in **Figure 2**.
The current is measured when different voltages are applied across component **X**.

Figure 2

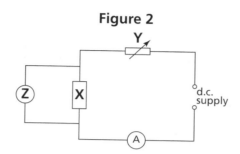

05.1 Name the components labeled **Y** and **Z** in the circuit.

Y = ... **[1 mark]**

Z = ... **[1 mark]**

05.2 What is the role of component **Y** in the circuit?

.... **[1 mark]**

Table 2 shows the measurements obtained in this experiment.

Table 2

Voltage (V)	−0.6	−0.4	−0.2	0	0.2	0.4	0.6	0.8
Current (mA)	0	0	0	0	0	50	100	150

05.3 Name the independent variable in this experiment.

Independent variable = ... **[1 mark]**

05.4 Name the dependent variable in this experiment.

Dependent variable = ... **[1 mark]**

Question 5 continues on the next page

05.5 Plot a graph on the axes in **Figure 3** using the data from **Table 2**.

Figure 3

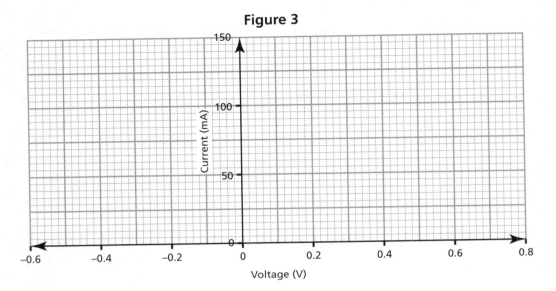

[3 marks]

05.6 The student looks at their measurements and decides that there are no anomalous results.

Are they correct?
You must explain your answer.

_____

_____ [1 mark]

05.7 Use the shape of the graph to name component **X**.

Component **X** = _____ [1 mark]

06   There are many isotopes of the element strontium (Sr).

06.1   What do the nuclei of different strontium isotopes have in common?

_____   **[1 mark]**

06.2   When the nucleus of a strontium-90 atom decays, it emits radiation and changes into a nucleus of yttrium-90

$$^{90}_{38}\text{Sr} \rightarrow \ ^{90}_{39}\text{Y} + \text{radiation}$$

What type of decay is this?

Answer _____   **[1 mark]**

06.3   Give a reason for your answer to **06.2**.

_____   **[1 mark]**

Strontium-90 has a half-life of 30 years.

06.4   What is meant by the term 'half-life'?

_____   **[1 mark]**

06.5   After 300 years, a sample of strontium-90 would be considered safe.

Explain why (you do not need to include calculations in your answer).

_____

_____

_____   **[2 marks]**

Turn over for the next question

07 A car that is moving has kinetic energy.
The faster a car goes, the more kinetic energy it has.

07.1 The mass of a car is 1050 kg.

Calculate the kinetic energy when the car is travelling at 30 m/s.
Give the unit.
Use the correct formula from the physics data sheet.

Kinetic energy = _____   **[4 marks]**

There is a government road safety campaign to reduce the speed at which people drive
in residential areas.
It uses the slogan 'Kill your speed, not a child'.
The scientific reason for this is that kinetic energy is transferred from the vehicle to the
person it knocks down.

07.2 A bus and car are travelling at the same speed.

Use the idea of kinetic energy to explain why the bus is likely to cause more harm
to a person who is knocked down than the car would.

_____

_____

_____   **[2 marks]**

07.3 A car and its passengers have a total mass of 1200 kg.
At 8 m/s the car has a kinetic energy of 38 400 J.

Calculate the increase in kinetic energy when the car increases its speed to 10 m/s.

Increase = _____ **J**   **[3 marks]**

07.4 Explain why the increase in kinetic energy is much greater than the increase in speed.

_____

_____   **[1 mark]**

Turn over for the next question

08 In order to jump over the bar, a high jumper must raise his mass above the ground by 1.25 m. The high jumper has a mass of 65 kg.

The gravitational field strength is 10 N/kg.

08.1 The high jumper just clears the bar.

Calculate the gain in his gravitational potential energy.

Gain = _____ J **[2 marks]**

08.2 A different higher jumper needs to gain 1000 J of gravitational energy to clear the same height.

Which of the following best describes the high jumper's mass?

Tick **one** box.

His mass is less than 65 kg. ☐

His mass is 65 kg. ☐

His mass is more than 65 kg. ☐ **[1 mark]**

08.3 When the high jumper jumps over the bar, kinetic energy is converted to gravitational energy.
The amount of kinetic energy is more than the gravitational energy gained.

Give **two** reasons why.

_____

_____

[2 marks]

09 The circuit diagram in **Figure 4** shows a circuit used to supply electricity for car headlights.

Figure 4

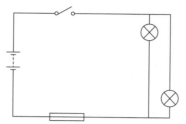

The current through the filament of one car headlight is 2 A.
The potential difference supplied by the battery is 12 V.

09.1 What is the potential difference across each headlight?

Potential difference = _____ V [1 mark]

09.2 Work out the total current through the battery.

Current = _____ A [1 mark]

09.3 Calculate the resistance of each headlight filament when in use.

Resistance = _____ Ω [2 marks]

Question 9 continues on the next page

09.4 How does the total resistance of the circuit compare to the resistance of each individual bulb?

_____ **[1 mark]**

09.5 Calculate the power supplied to each of the two headlights of the car.

Power = _____ **W** **[2 marks]**

09.6 The fully charged car battery can deliver 96 kJ of energy at 12 V.

How long can the battery keep both the headlights fully on?

Length of time = _____ **s** **[2 marks]**

END OF QUESTIONS

Collins

GCSE
COMBINED SCIENCE
Physics: Paper 2 Foundation Tier

Materials

Time allowed: 1 hour 15 minutes

For this paper you must have:
- a ruler
- a calculator
- a protractor
- the Physics Equation Sheet (page 167).

Instructions

- Answer **all** questions in the spaces provided.
- Do all rough work on the page. Cross through any work you do not want to be marked.

Information

- There are **70** marks available on this paper.
- The marks for each question are shown in brackets [ ].
- You are expected to use a calculator where appropriate.
- You are reminded of the need for good English and clear presentation in your answers.
- When answering questions 04.1 and 06.5 you need to make sure that your answer:
  - is clear, logical, sensibly structured
  - fully meets the requirements of the question
  - shows that each separate point or step supports the overall answer.

Advice

- In all calculations, show clearly how you work out your answer.

01   A student used a lever system to investigate how the force of attraction between a coil and an iron rocker varied with the current in the coil.
     She supported a coil vertically and connected it to an electrical circuit as shown in **Figure 1**.
     The weight of the iron rocker is negligible.

Figure 1

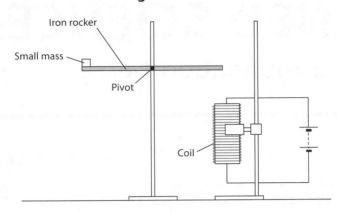

01.1 Why is it important that the rocker in this experiment is made of iron?

_____

[1 mark]

01.2 The student put a small mass on the end of the rocker and adjusted the current in the coil until the rocker balanced.

To keep the rocker balanced, how will the current through the coil need to change as the size of the mass is increased?

_____

[1 mark]

01.3 Explain your answer to **01.2**.

_____

_____

_____

[2 marks]

01.4 A second student set up the same experiment and put an iron core inside the coil.

How will this affect the strength of the magnet and, therefore, the size of the mass that can be balanced?

_____

_____

_____ **[2 marks]**

Turn over for the next question

02   A group of students investigate the motion of different objects.
They study the movement of cars on a racing track.
They also study the orbit of the Moon around Earth.

02.1   Each lap of the racing track is 1.2 miles long.

Calculate the total distance travelled by a car after five complete laps.

Distance = _____   [1 mark]

02.2   The displacement after five complete laps is zero. Explain why.

_____

_____   [2 marks]

02.3   On one of the bends, the speed of the cars is unchanged, so the students conclude that the resultant force is zero.

Are they correct? Explain your answer.

_____

_____   [1 mark]

02.4   The Moon orbits the Earth in a circular path.

direction	resistance	speed	velocity

Use words from the box to complete the sentences.
You may use each word once, more than once or not at all.

The Moon's _____ is constant but its _____ changes.

This is because its _____ changes.   [3 marks]

02.5   What provides the force needed to keep the Moon in its orbit around the Earth?

_____   [1 mark]

03 Radio waves, ultraviolet waves, visible light and X-rays are all types of electromagnetic radiation.

03.1 Choose wavelengths from the list below to complete **Table 1**.

3×10^{-8} m 1×10^{-11} m 5×10^{-7} m

Table 1

Type of radiation	Wavelength (m)
Radio waves	1500 m
Visible light	
Ultraviolet waves	
X-rays	

[2 marks]

03.2 From the list below select **two** properties that are common to all the waves in **Table 1**.

Tick **two** boxes.

They all travel at the same speed in a vacuum.   ☐

They all have the same frequency in a vacuum.   ☐

They are all longitudinal waves.   ☐

They are all transverse waves.   ☐

[2 marks]

03.3 State the name and a use of **one** type of electromagnetic wave **not** listed in **Table 1**.

[2 marks]

Question 3 continues on the next page

03.4 Radio waves can be used to control remote control cars.

Calculate the frequency of radio waves of wavelength 300m. Give the unit.
(The velocity of electromagnetic waves is 3×10^8 m/s.)

Frequency = _____ [4 marks]

The graph in **Figure 2** shows the speed of a remote-controlled vehicle during a race.

Figure 2

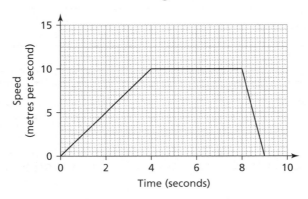

03.5 Calculate the acceleration during the first four seconds.

Acceleration = _____ m/s² [3 marks]

03.6 What is the maximum speed reached by the vehicle?

Maximum speed = _____ m/s [1 mark]

03.7 Between 4 and 8 seconds the speed of the remote control car is 10 m/s.

Calculate the distance it travels during this time.

Distance = _____ m [2 marks]

04 The distance–time graph in **Figure 3** represents the motion of a car during a race.

Figure 3

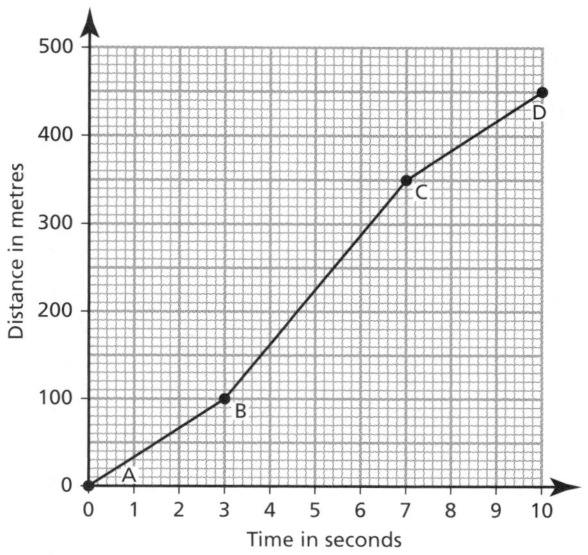

04.1 Describe the motion of the car between points **A** and **D**.

Refer to the speed of the car and the points where the car changes speed as shown on the graph in **Figure 3**.
You should calculate the maximum speed (between points **B**–**C**) reached by the car and use this in your answer.

....

....

....

....

....

....

....

[6 marks]

Question 4 continues on the next page

04.2 At the start of the race, the car accelerates from rest to a speed of 30 m/s in 6 seconds.

Calculate the acceleration of the car.

Acceleration = _____ m/s^2 **[3 marks]**

05   A car of mass 1200 kg has an engine thrust of 3500 N and experiences a resistive force
      of 2000 N.

    05.1   Calculate the acceleration of the car.
            Work out the resultant force on the car and use this to calculate the acceleration.

                                    Acceleration = ... m/s²   [4 marks]

    05.2   Use the idea of terminal velocity to explain why the car reaches a top speed even
            though the thrust force remains at 3500 N.

            ....

            ....

            ....   [2 marks]

The driver of the car notices a hazard.
They apply the brakes to come to a complete stop in a certain distance.
This stopping distance is made up of the thinking distance and the braking distance.

    05.3   What is meant by the term 'thinking distance'?

            ....

            ....   [1 mark]

    05.4   State two factors that affect thinking distance.

            ....

            ....

            ....   [2 marks]

Question 5 continues on the next page

05.5 The braking distance of a car depends on the speed of the car and the braking force applied.

State **one** other factor that affects the braking distance.

....

.... **[1 mark]**

05.6 During braking, the temperatures of the brakes increase.

Explain why.

....

....

....

.... **[2 marks]**

06   **Figure 4** shows two waves.

Figure 4

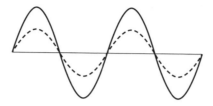

06.1   Name **one** wave quantity that is the same for both waves.

.... [1 mark]

06.2   Name **one** wave quantity that is different for the two waves.

.... [1 mark]

06.3   The waves shown in **Figure 4** are transverse waves.

Which of the following types of wave is **not** a transverse wave?
Draw a ring around the correct answer.

gamma rays   **sound**   **visible light**   [1 mark]

06.4   A student studies waves in a ripple tank.
Every second, eight waves pass a fixed point in the tank.
The waves have a wavelength of 0.015 m.

Calculate the speed of the water waves.

Wave speed = ... m/s   **[3 marks]**

Question 6 continues on the next page

06.5 Measuring the wavelength of moving water waves is difficult.
To improve accuracy, the student decides to use a stroboscope and a ruler.

Describe the procedure that the student should use and outline **one** hazard they need to be aware of.

[4 marks]

07   **Figure 5** shows two lines of the magnetic field pattern around a current-carrying wire.

Figure 5

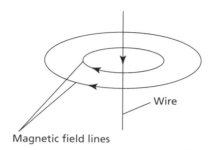

Magnetic field lines

Wire

07.1   The direction of the current is reversed.

How does this affect the lines in the magnetic field pattern?

.... **[1 mark]**

07.2   The size of the current through the wire is increased.

How does this affect the lines in the magnetic field pattern?

.... **[1 mark]**

07.3   The wire is coiled into a solenoid.

Draw a solenoid and the magnetic field around and through the solenoid.

[3 marks]

Question 7 continues on the next page

07.4 What effect does coiling the wire into a solenoid have on the strength of the magnetic field?

_____ **[1 mark]**

07.5 A group of students uses a plotting compass to study the magnetic field produced by the solenoid.
They place the plotting compass at different positions and record how strongly it is affected when the solenoid is switched on.

Sort the following positions in to order, starting with the strongest effect and ending with the weakest.

A Compass 3 cm from the solenoid, placed in line with the end.

B Compass 6 cm from the solenoid, placed at the side.

C Compass inside the solenoid.

D Compass 6 cm from the solenoid, placed in line with the end.

_____ **[3 marks]**

END OF QUESTIONS

Physics Equation Sheet

1	(final velocity)² – (initial velocity)² = 2 × acceleration × distance	$v^2 - u^2 = 2as$
2	elastic potential energy = 0.5 × spring constant × (extension)²	$E_e = \frac{1}{2}ke^2$
3	change in thermal energy = mass × specific heat capacity × temperature change	$\Delta E = mc\Delta\theta$
4	period = $\dfrac{1}{\text{frequency}}$	
5	thermal energy for a change of state = mass × specific latent heat	$E = mL$

The Periodic Table

Key

relative atomic mass
atomic symbol
name
atomic (proton) number

Example:

1
H
hydrogen
1

Group 1	Group 2																Group 3	Group 4	Group 5	Group 6	Group 7	Group 0
																						4 **He** helium 2
7 **Li** lithium 3	9 **Be** beryllium 4																11 **B** boron 5	12 **C** carbon 6	14 **N** nitrogen 7	16 **O** oxygen 8	19 **F** fluorine 9	20 **Ne** neon 10
23 **Na** sodium 11	24 **Mg** magnesium 12																27 **Al** aluminium 13	28 **Si** silicon 14	31 **P** phosphorus 15	32 **S** sulfur 16	35.5 **Cl** chlorine 17	40 **Ar** argon 18
39 **K** potassium 19	40 **Ca** calcium 20	45 **Sc** scandium 21	48 **Ti** titanium 22	51 **V** vanadium 23	52 **Cr** chromium 24	55 **Mn** manganese 25	56 **Fe** iron 26	59 **Co** cobalt 27	59 **Ni** nickel 28	63.5 **Cu** copper 29	65 **Zn** zinc 30						70 **Ga** gallium 31	73 **Ge** germanium 32	75 **As** arsenic 33	79 **Se** selenium 34	80 **Br** bromine 35	84 **Kr** krypton 36
85 **Rb** rubidium 37	88 **Sr** strontium 38	89 **Y** yttrium 39	91 **Zr** zirconium 40	93 **Nb** niobium 41	96 **Mo** molybdenum 42	[98] **Tc** technetium 43	101 **Ru** ruthenium 44	103 **Rh** rhodium 45	106 **Pd** palladium 46	108 **Ag** silver 47	112 **Cd** cadmium 48						115 **In** indium 49	119 **Sn** tin 50	122 **Sb** antimony 51	128 **Te** tellurium 52	127 **I** iodine 53	131 **Xe** xenon 54
133 **Cs** caesium 55	137 **Ba** barium 56	139 **La*** lanthanum 57	178 **Hf** hafnium 72	181 **Ta** tantalum 73	184 **W** tungsten 74	186 **Re** rhenium 75	190 **Os** osmium 76	192 **Ir** iridium 77	195 **Pt** platinum 78	197 **Au** gold 79	201 **Hg** mercury 80						204 **Tl** thallium 81	207 **Pb** lead 82	209 **Bi** bismuth 83	[209] **Po** polonium 84	[210] **At** astatine 85	[222] **Rn** radon 86
[223] **Fr** francium 87	[226] **Ra** radium 88	[227] **Ac*** actinium 89	[261] **Rf** rutherfordium 104	[262] **Db** dubnium 105	[266] **Sg** seaborgium 106	[264] **Bh** bohrium 107	[277] **Hs** hassium 108	[268] **Mt** meitnerium 109	[271] **Ds** darmstadtium 110	[272] **Rg** roentgenium 111	[285] **Cn** copernicium 112						[286] **Uut** ununtrium 113	[289] **Fl** flerovium 114	[289] **Uup** ununpentium 115	[293] **Lv** livermorium 116	[294] **Uus** ununseptium 117	[294] **Uuo** ununoctium 118

* The Lanthanides (atomic numbers 58–71) and the Actinides (atomic numbers 90–103) have been omitted.

Relative atomic masses for Cu and Cl have not been rounded to the nearest whole number.

Answers

Topic-Based Questions

Page 6 Cell Structure

1. a) B   [1]
   b) A   [1]
   c) C   [1]
2. a) B [1]; it has a nucleus / chloroplasts [1]
   b) Eukaryotic cells are much larger than prokaryotic [1]
   c) Using flagella [1]; which whip around [1]

Page 7 Investigating Cells

1. a) To act as a stain [1]; because some of the structures are transparent [1]
   b) i) 30 micrometres   [1]
      ii) $\frac{20}{0.03}$ [1]; = 667, magnified 667 times (×667) [1]

2.

	Light Microscope	Electron Microscope
Can be used to see both animal and plant cells.	✓	✓
Works by passing light through the specimen.	✓	✗
Can be used to see the detailed structure of mitochondria.	✗	✓

(1 mark for each correct row)   [3]

Page 8 Cell Division

1. a) Mitosis   [1]
   b) Chromosomes are moving to the ends of the cell [1]; pulled by spindles [1]
2. a) Stem cells can divide to produce different types of cells [1]; these cells can replace lost or defective cells or be grown into tissues [1]
   b) **Any one of:** the stem cells may come from cloned embryos [1]; embryos may be destroyed in the process [1]; concerned that the long-term effects are not known [1]

Page 9 Transport In and Out of Cells

1. a) 6 × (10 × 10) = 600   [1]
   b) Diffusion   [1]
   c) Block C [1]; it has the largest surface area [1]; so there is more surface for the dye to diffuse across [1]
2. a) Osmosis always involves movement of water particles, whereas diffusion is movement of particles [1]; osmosis involves a cell membrane and diffusion does not [1]
   b) Active transport is movement of particles against a concentration gradient / from low to high concentration, rather than from high to low concentration as with diffusion [1]; active transport requires energy from respiration (diffusion does not require an input of energy from the cell) [1]

Page 10 Levels of Organisation

1. a) cells, tissues, organs, systems   [1]
   b) **Top to bottom:** organ [1]; cell [1]; organ [1]; tissue [1]
   c) Muscle cells are specialised for contraction.   [1]
2. lignin [1]; water [1]; holes [1]

Page 11 Digestion

1. a) The rate of reaction increases to a peak / optimum [1]; and then decreases [1]
   b) 40°C   [1]
2. a) Pancreas / small intestine   [1]
   b) Olestra molecules are the wrong shape.   [1]

Page 12 Blood and the Circulation

1. **Top to bottom:** 3, 1, 2, 4 [3] (2 marks for two correct; 1 mark for one correct)
2. a) Haemoglobin combines with oxygen at the lungs, when in high concentration [1]; and forms oxyhaemoglobin [1]; it releases oxygen at the tissues, as the concentration is low there, to go back to haemoglobin [1]

   > haemoglobin + oxygen ⇌ oxyhaemoglobin

   b) i) Plasma   [1]
      ii) Lungs   [1]

Page 13 Non-Communicable Diseases

1. a) i) The heart stops beating (Accept an accurate explanation of why)   [1]
      ii) It restricts blood flow to the heart [1]; so not enough oxygen and glucose are supplied to cells in the heart muscle for respiration [1]
   b) Having an atheroma does not seem to depend on level of physical activity [1]; increased exercise reduced incidence of heart disease [1]; suggesting atheromas are not the only factor in causing heart disease [1]

Page 14 Transport in Plants

1. a) water [1]; transpiration [1]; windy [1]
   b) i) A smaller change in mass due to less transpiration [1]; because less water is taken up by the roots [1]
      ii) A smaller change in mass due to less transpiration [1]; as some of the stomata have been blocked [1]
2. **Any two of:** water is transported through the xylem, sugars are transported in phloem [1]; water is moved upwards only, sugars move all over [1]; water movement does not need energy from respiration, sugars move by active transport [1]

Page 15 Pathogens and Disease

1. a) Four correctly drawn lines [3] (2 marks for two lines; 1 mark for one line)
      rose black spot – fungus
      salmonella – bacterium
      measles – virus
      malaria – protozoan
   b) vector [1]; *Plasmodium* [1]; host [1]
2. a) **Any two of:** infected needles [1]; sexual activity [1]; across the placenta [1]
   b) The virus attacks white blood cells [1]; making the immune system less effective [1]
3. To prevent the spread of the disease [1]; to destroy the spores [1]

Page 16 Human Defences Against Disease

1. Nose: hairs trap particles [1]; skin: sebum kills bacteria [1]; stomach: acid kills microorganisms [1]
2. A = antigen [1]; B = antibody [1]; C = phagocyte [1]

Page 17 Treating Diseases

1. a) HIV is a virus [1]; TB is a bacterial infection [1]
   b) Many TB bacteria are resistant [1]; and are not killed by antibiotics [1]
   c) **Any two of:** finish the dose [1]; only prescribe if necessary [1]; rotate antibiotics used [1]
   d) **Any two of:** to test if they are effective / work [1]; to see if they have any side effects / are safe [1]; to work out the appropriate dosage [1]

Page 18 Photosynthesis

1. a) carbon dioxide + water [1]; ⟶ glucose + oxygen [1]
   b) i) **Any two of:** root tip [1]; shoot tip [1]; fruits [1]; seeds [1]; storage organs [1];
      ii) **Any two of:** starch [1]; for storage [1] OR lipids [1]; for storage [1] OR proteins [1]; for growth [1] OR cellulose [1]; for cell walls [1]

Answers

2. Some of the mass did come from the water [1]; but some also comes from carbon dioxide / photosynthesis [1]

Page 19 Respiration and Exercise

1. a) oxygen [1]; carbon dioxide [1]; water [1] (The two products can be given in any order)
   b) i) 1.4 (mols per litre) [1]
      ii) In the race / after training, lactic acid does not start to increase so early [1]; and does not increase as much [1]; lactic acid causes muscle cramps, and because there is less she can run more efficiently [1]

Page 20 Homeostasis and the Nervous System

1. a) **Any two of:** rapid [1]; protects the body [1]; does not need thought [1]
   b) A receptor [1]
2. a) A (sensory) neurone [1]
   b) Transmitter molecules are released when an impulse reaches a synapse [1]; they diffuse across the synapse [1]; and stimulate an impulse in the next neurone [1]

Page 21 Hormones and Homeostasis

1. endocrine [1]; pituitary [1]; blood [1]
2. a) Type 1 [1]
   b) Insulin [1]
   c) **Any two of:** blood glucose level too high [1]; glucose starts to pass out in urine [1]; coma and / or death [1]
   d) i) Type 2 [1]
      ii) Treat by modifying diet [1]; rather than by insulin injections [1]

Page 22 Hormones and Reproduction

1. a) Hormone A: Oestrogen [1]; inhibit FSH release / stimulates LH release / repairs lining of uterus [1]; Hormone B: progesterone [1]; maintains the lining of the uterus / inhibits FSH and LH [1]
   b) i) X marked just after peak of Hormone A [1]
      ii) Y marked at start or end of the cycle [1]
   c) i) Testosterone [1]
      ii) Testes [1]
2. The pill contains both oestrogen and progesterone [1]; it inhibits FSH [1]; so that eggs do not develop / no ovulation [1]

Page 23 Sexual and Asexual Reproduction

1. a) It produces runners [1]; that touch the ground and root, growing into new plants [1]
   b) **Any two of:** it produces flowers [1]; pollen is transferred from plant to plant [1]; it makes seeds [1]
2. a) chromosomes [1]; DNA [1]; gene [1]
   b) To help to identify / understand / treat genetic disorders **OR** to study evolution [1]

Page 24 Patterns of Inheritance

1. a) i) Sachin / Rose [1]
      ii) Sara / Tia / Rohit [1]
   b) Sachin: gg (genotype) [1]; male with Gaucher disease (phenotype) [1]; Tia: female with normal enzyme (phenotype) [1]
   c) Probability: 50% / $\frac{1}{2}$ / 1 : 1 [1]
   d) **Any two of:** She might prefer not know until it is born [1]; to avoid having to decide whether to have a termination [1]; the test may increase the risk of miscarriage [1] (Accept any other sensible reason)

Page 25 Variation and Evolution

1. Genetic: Bill and Ben have brown eyes [1]; Environment: Ben has a scar [1]; Genetics and Environment: Bill is 160cm tall [1]; Ben's body mass is 60kg [1]
2. a) Darken / turn black [1]
   b) The dark moths are better camouflaged [1]; so fewer are eaten by birds [1]
   c) Natural selection [1]

Page 26 Manipulating Genes

1. Selective breeding [1]; genetic engineering [1]
2. a) i) Elvira [1]
      ii) Because it grows upright [1]
   b) i) Elvira and calypso [1]
      ii) Transfer pollen between them [1]; grow the seeds produced and select the plants that best show the characteristics wanted [1]; repeat many times [1]

Page 27 Classification

1. a) Organisms that can mate with each other [1]; to produce fertile offspring [1]
   b) i) Bobcat [1]; ocelot [1]
      ii) They are both in the same genus [1]; *Felis* [1]
2. a) No more left alive [1]
   b) Disturbed them, so they did not breed [1]
   c) Study fossils [1]

Page 28 Ecosystems

1. a) i) To spread their weight / stop them sinking in the sand [1]
      ii) To shade them from the sun / store of food or fat / produce water from respiration (by breaking down the fat store) [1]
   b) population [1]; habitat [1]; community [1]; competed [1]; biotic [1]
   c) i) Use a quadrat [1]; placed at random [1]; count cacti present in quadrat [1]; repeat many times [1]; calculate the mean number per m² [1]; multiply results up by the area [1]
      ii) Camels move around [1]; so may count them more than once / not at all [1]

Page 29 Cycles and Feeding Relationships

1. A = Photosynthesis [1]; B = Respiration [1]; C = Decomposition [1]; D = Combustion / burning [1]; E = Eating / feeding [1]
2. a) Predator–prey graph [1]
   b) Secondary consumer [1]
   c) Rabbit numbers increase, so there so more food available for the foxes [1]; more foxes survive and numbers increase [1]; this means more rabbits are eaten, so the fox numbers drop again (as less food is available) [1]

Page 30 Disrupting Ecosystems

1. a) There is no pollution / no industry / no cars [1]
   b) **Any two of:** in summer, there is more sunlight [1]; so more photosynthesis [1]; and more carbon dioxide taken in by plants [1]
   c) Greenhouse effect / greenhouse gas [1]; allows heat from the sun through the atmosphere [1]; but absorbs / traps energy reflected back from the Earth's surface, leading to a rise in temperature [1]
2. a) Frog [1]
   b) When some fossil fuels are burned [1]; acidic gases are released, e.g. sulfur dioxide gas [1]; which dissolves in water in the atmosphere and then falls as acid rain [1]

Page 31 Atoms, Elements, Compounds and Mixtures

1. a) $2Cu + O_2 \rightarrow 2CuO$ [1]
   b) 18.9 − 15.6 = 3.3g [1]
   c) 0.1g [1]

> The resolution of the balance is the degree of accuracy. In this case, the measurements are given to the nearest 0.1g.

2. a) Place the salt solution in the round bottomed flask [1]; heat the solution [1]; and collect the liquid that distils at 100°C [1]

   b) **Any one of:** Burns from the Bunsen burner / hot water / steam [1]; equipment may crack so risk of cuts [1]

Page 32 Atoms and the Periodic Table

1. a) Thomson showed that the atom could be divided up into simpler substances [1]; the old model was no longer correct [1]

   b) Electron [1]

2. a) Based on Thomson's model, the alpha particles would have all gone through [1] OR There would have been no deflection [1] (Accept: It went against Thomson's model)

   b) To reduce the effects of errors [1]; to check repeatability [1] (Accept: To make sure they were correct / accurate for 1 mark)

   c) **Any three of:** small nucleus [1]; nucleus in the centre of the atom [1]; nucleus has a positive charge [1]; electrons in orbit around the nucleus [1] (Accept a labelled diagram)

Page 33 The Periodic Table

1. C [1]

> The number of electrons in the outer shell is the same as the group that the element is in. The number of electron shells is the same as the row (period) that the element is in.

2. a) X [1]; because it does not conduct electricity [1]

   b) W [1]; because it is a liquid at room temperature [1]; and it conducts electricity [1]

   c) (lithium) hydroxide [1]; hydrogen [1]

Page 34 States of Matter

1. a) g [1]; s [1]

> Gas = g, solid = s, liquid = l, aqueous / dissolved in water = aq

   b) 650°C [1]
   c) −183°C [1]
   d) **Any six of:** Oxygen has a low boiling point [1]; magnesium oxide has a high boiling point [1]; oxygen has weak forces of attraction between its particles / molecules [1]; magnesium has strong forces in between its ions / particles [1]; oxygen is a simple molecule [1]; magnesium oxide is an ionic compound [1]; only a small amount of energy is needed to boil oxygen / turn it from a liquid into a gas [1]; a lot of energy is needed

to boil magnesium oxide / turn it from a liquid into a gas [1]

Page 35 Ionic Compounds

1. a) calcium chloride ($CaCl_2$) [1]; sodium carbonate (Na_2CO_3) [1]

> Ionic compounds contain both metal and non-metal elements. Look at the periodic table if you are not sure whether an element is a metal or a non-metal. The metals are on the left side, the non-metals on the right.

   b) They cannot conduct electricity when solid because the ions cannot move [1]; they can conduct electricity when in solution because the ions are free to move about [1]; and carry the charge [1]

2. Electrons are transferred from magnesium to bromine [1]; the magnesium atom loses two electrons [1]; forming Mg^{2+} / 2+ ions [1]; the two bromine atoms each gain one electron [1]; forming Br^- / 1− ions [1]

Page 36 Covalent Compounds

1. The forces between bromine molecules are stronger [1]

2. a) A shared pair of electrons drawn between H and Br [1]; no additional hydrogen electrons and three non-bonding pairs shown on bromine [1] (second mark dependent on first)

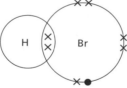

   b) HBr [1]

3. High melting point – Strong covalent bonds between many carbon atoms [1]; Does not conduct electricity when molten – There are no charged particles that are free to move [1]

> The covalent bonds between atoms are very strong. The force of attraction between molecules is called the intermolecular force and is much weaker.

Page 37 Metals and Special Materials

1. High melting point: strong force of attraction between positive ions and negative electrons / metallic bond is strong [1]; so, a lot of energy is needed to break metallic bonds (to melt metal) [1]; thermal conductivity: electrons are delocalised [1]; and are free to move through the metal and transfer energy [1]; malleability: the

ions are arranged in layers / have a regular arrangement [1]; the layers are able to slide over each other easily [1]

Page 38 Conservation of Mass

1. a) $Zn(s) + 2HCl(aq) \rightarrow ZnCl_2(aq) + H_2(g)$ [1]

   b) Gas is produced in the flask (increasing the pressure inside) [1]; the flask may crack / the bung may be forced out [1] (Accept: 'it is not safe' for 1 mark)

   c) The mass reading will decrease [1]; because the hydrogen particles (atoms) in the hydrochloric acid are rearranged [1]; to form hydrogen gas, which leaves the flask into the air [1]

Page 39 Reactivity of Metals

1. a) Magnesium [1]; sodium [1]
   b) It is cheaper / more abundant [1]
2. Displacement [1]

Page 40 The pH Scale and Salts

1. 4 [1]
2. a) Add excess copper(II) oxide to acid (accept alternatives, e.g. 'until no more will react') [1]; filter (to remove excess copper(II) oxide) [1]; heat filtrate to evaporate some water or heat to point of crystallisation [1]; leave to cool (so crystals form) [1]

   b) **Any one of:** wear apron [1]; use eye protection [1]; tie hair back [1]

   c) Copper(II) chloride [1]

Page 41 Electrolysis

1. a) Aluminium is more reactive than carbon [1]
   b) i) Oxygen / O_2 [1]
      ii) Aluminium / Al [1]
   c) It requires a lot of electricity to melt the aluminium oxide / keep it molten [1]; so aluminium oxide is dissolved in molten cryolite [1]; to reduce the melting point, so less electricity is used [1]

Page 42 Exothermic and Endothermic Reactions

1. a) It reduces the movement of heat to and from the surroundings [1]; which could affect the accuracy of the results [1] (Accept 'it is an insulator' for 1 mark)

   b) **Any two of:** type of acid [1]; concentration of acid [1]; surface area of metal [1]; temperature of acid [1]; volume of acid [1]; mass of metal [1]

   c) The more reactive the metal [1]; the more exothermic the reaction [1]

Answers

Page 43 Rate of Reaction

1. a) Repeat using different concentrations of acid [1]; for example, 0.5mol/dm³, 1mol/dm³, 1.5mol/dm³, 2mol/dm³ [1] (Accept any other concentrations within a sensible range)
   b) **Any one of:** wear eye protection [1]; wear apron [1]; do not heat mixture over 50°C [1]
   c) The time taken for the cross to disappear will decrease as concentration of acid increases [1]; as the (acid) concentration increases so does the number of (acid) particles in a given volume [1]; so they collide more often with the sodium thiosulfate particles [1]; resulting in more successful collisions and a reaction taking place (sulfur forming) [1]

Page 44 Reversible Reactions

1. a) Reversible reaction / equilibrium [1]
   b) Forward reaction [1]; the backward reaction is endothermic [1]; because heat is needed to decompose the ammonium chloride [1]
   c) The forward and reverse reactions are taking place at the same rate. [1]; The amounts of reactants and products are constant. [1]

Page 45 Alkanes

1. a) A molecule / compound [1]; that only contains carbon and hydrogen [1]
   b) The different molecules / hydrocarbons condense [1]; at a place in the column just below their boiling point [1]
2. O_2 [1]; 2 [1]

Page 46 Cracking Hydrocarbons

1. a) 2 [1]
   b) C_8H_{18} only [1]
2. Add bromine water [1]; if it goes colourless it is an alkene [1]; if it stays orange / brown / does not change colour then it is an alkane [1]

Page 47 Chemical Analysis

1. a) Paper [1]
   b) Pencil [1]; the ink from the pen would 'run' on the paper / dissolve in the solvent and affect the results [1]
   c) B and C [1]; there is only one spot [1]

d) 18cm (distance moved by X) and 28cm (distance moved by solvent) [1]; $\frac{18}{28}$ [1] = 0.64286 = 0.64 to 2 d.p. [1] (award 2 marks if calculation is correct but the answer is not given to 2 d.p.)

Page 48 The Earth's Atmosphere

1. C [1]
2. $\frac{0.05}{4.2} \times 100$ [1]; = 1.2% [1]
3. a) Evidence [1]; that supported their theory [1]
   b) **Any three of:** it happened a long time ago [1]; nobody was alive to record how it happened [1]; there is little evidence available [1]; there is evidence to support both theories [1]

Page 49 Greenhouse Gases

1. a) Carbon dioxide OR methane [1]; **plus:** increased burning of fossil fuels / increased deforestation (for carbon dioxide) OR more animal farming (digestion, waste decomposition) / decomposition of rubbish in landfill sites (for methane) [1]
   b) From: 0.6°C [1]; To: 3.6°C [1] (If no units are included only award 1 mark)
   c) **Any two of:** complex systems [1]; many different variables [1]; may only be based on parts of evidence [1]

Page 50 Earth's Resources

1. a) Water that is safe to drink [1]
   b) It contains dissolved substances [1]; it is a mixture as it contains more than just water molecules [1]
   c) **Any one of:** adding chlorine [1]; adding ozone [1]; using UV light [1]
   d) To remove the water from the salt [1]
   e) The sea water needs to be heated [1]; which requires a large amount of energy [1]

Page 51 Using Resources

1. **Any six of:** wood pulp is from trees, a renewable resource [1]; trees should be replanted before wood pulp can be considered a sustainable resource [1]; clay, chalk and titanium oxide are quarried which can have negative environmental impacts [1]; paper production uses a lot of water [1]; transportation of plastic bags uses less fuel [1]; plastic is longer lasting / can be reused many times [1]; (plastic longer lasting) so plastic may use less finite resources [1]; (plastic may use less finite resources) and so plastic

may have a lower energy requirement [1]; paper is biodegradable so spends less time in landfill [1]; paper is more likely to be recycled which lowers raw material usage [1]

> For this type of question, the examiner will expect your answer to be given in a clear and logical way, using good English and correct grammar and punctuation. They will award 5–6 marks for a clear description of the advantages and disadvantages of both types of bag, with logical links; 3–4 marks if a number of relevant points are made, but the logic is unclear; 1–2 marks for fragmented points, with no logical structure.

Page 52 Forces – An Introduction

1. a) Weight / gravity [1]; non-contact [1]; air resistance (or drag) [1]; contact [1]
   b) weight = mass × gravitational field strength / $W = mg$ [1]; weight = 120 000 × 10 = 1 200 000N [1] (Accept: 1.2×10^6)
   c) The lift force will be the same as the weight. [1]
   d) Constant [1]
2. a) Friction [1]
   b) Friction is a contact force [1]; lifting the glider off the track means there is no contact, so no friction [1]

Page 53 Forces in Action

1. a) Remove the weights [1]; if it returns to its original shape it was behaving elastically [1]
   b) i) 8.0 (cm) [1]
      ii) The extension appears to be linear [1]; and is increasing by 4cm each time [1]
   c) force [1]; 0N [1]; 6N [1]
   d) It allows you to check for errors / anomalies [1]; you can calculate a mean (average) result [1]

Page 54 Forces and Motion

1. a) Zero [1]

> Displacement is the distance from the start point. The person returns home – it is a circular journey – so the total displacement at the end of their journey is zero.

   b) B [1]
   c) Stationary [1]
   d) A is travelling away from home [1]; D is travelling in the opposite direction (back towards home) [1]; D is travelling slightly faster than A [1]

2. a) speed = distance travelled / time

   $v = \frac{s}{t}$ **[1]**; $= \frac{180}{6}$ = 30m/s **[1]**

   b) i) 50 − 40 = 10m/s **[1]**
      ii) 50 + 40 = 90m/s **[1]**

   > The velocity changes from 50m/s to the east to 40m/s to the west.

Page 55 Forces and Acceleration

1. a) The mass of the system **[1]**
   b) The force accelerating the trolley (provided by the hanging masses) **[1]**
   c) The acceleration of the trolley **[1]**

   > The independent variable is the one deliberately changed and the dependent variable is the one being measured.

2. a) acceleration = change in velocity / time taken

   $a = \frac{\Delta v}{t}$ **[1]**; acceleration = $\frac{0.05}{4.2} \times 100$
   = 1.5 **[1]**; m/s² **[1]**

   b) resultant force = mass × acceleration / $F = ma$ **[1]**; force = 68 × 1.5 = 102N **[1]**

3. a) acceleration = $\frac{5.2}{79.5}$ **[1]**; = 3m/s² **[1]**

   b) force = 950 × 3 **[1]**; = 2850N **[1]**

Page 56 Terminal Velocity, Stopping and Braking

1. They could change position e.g. dive head first **[1]**; to become more aerodynamic / reduce air resistance **[1]**
2. a) 1.4 seconds **[1]**
   b) 1.4 × 15 **[1]**; = 21m **[1]**
   c) 4 − 1.4 = 2.6 seconds **[1]**
   d) It would take longer to come to a stop **[1]**
   e) A correctly drawn graph line that starts horizontally at 15m/s **[1]**; then starts sloping downwards between 0.2s and 0.8s **[1]**; and has the same gradient on the downslope as the original line **[1]**

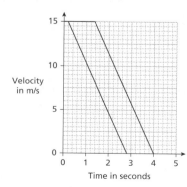

f) **Any two of:** The down slope would start at the same point **[1]**; but have a shallower gradient **[1]**; and take a longer total time to stop **[1]**

Page 57 Energy Stores and Transfers

1. a) 100 − 20 = 80 degree change **[1]**; energy = 2 × 4200 × 80 **[1]**; = 672 000J **[1]**
   b) 672 000 = 2.5 × 4200 × temp change **[1]**; temp change = $\frac{3}{25} \times 100$ = 64°C **[1]**; final temp = 20 + 64 = 84°C **[1]**
2. a) gravitational potential energy = mass × gravitational field strength × height / $E_p = mgh$ **[1]**; $E_p = 0.1 \times 10 \times 0.05$ **[1]**; = 0.05J **[1]**
   b) kinetic energy = 0.5 × mass × (speed)² / $E_k = \frac{1}{2}mv^2$ **[1]**; 0.05 = 0.5 × 0.1 × v^2 **[1]**; $v^2 = \frac{15}{22}$ = 1, v = 1m/s **[1]**

Page 58 Energy Transfers and Resources

1. light **[1]**; electrical / thermal **[1]**; heat **[1]**
2. a) Start temperature of water **[1]**; thickness of fleece **[1]**
   b) Fleece M **[1]**; because it insulates the best **[1]**

Page 59 Waves and Wave Properties

1. a) Half a wave per second **[1]** (Accept: 1 wave every 2 seconds)
   b) wave speed = frequency × wavelength / $v = f\lambda$ **[1]**
   c) speed = $\frac{distance}{time}$ / $v = \frac{s}{t}$, $v = \frac{50}{10}$ **[1]**; = 5m/s **[1]**
   d) $v = f\lambda$, $\lambda = \frac{5}{0.5}$ **[1]**; = 10m **[1]**
   e) Refraction **[1]**
2. In longitudinal waves, the particles oscillate **[1]**; parallel to the direction of energy transfer / wave motion **[1]**; in transverse waves, the oscillation is at right-angles to the direction of energy transfer / wave motion **[1]**

Page 60 Electromagnetic Waves

1. a) Can be shown on the diagram to help explain but must include 'refraction' in the answer for full marks, e.g.
   Light rays from the pin **[1]**; are refracted when they leave the water **[1]**; away from the normal and into the eye **[1]**

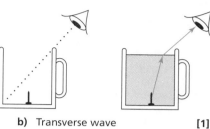

   b) Transverse wave **[1]**
2. frequency **[1]**; wavelength **[1]**

Page 61 The Electromagnetic Spectrum

1. a) Microwaves **[1]**
   b) Accept any sensible answer, e.g. X-rays for photographing bones OR gamma rays for sterilisation OR UV for sunbeds **[2]** (1 mark for wave, 1 mark for use)
2. a) X-rays **[1]**
   b) It can penetrate soft tissue **[1]**; but is blocked by bone **[1]**
3. a) Skin cancer **[1]**
   b) It has more / higher frequency energy **[1]**; and is ionising **[1]**
   c) **Any two of:** they think it looks good / healthy **[1]**; they don't think it will happen to them **[1]**; they don't think it is that risky **[1]** (Accept any other sensible answer)

Page 62 An Introduction to Electricity

1. current **[1]**; charge **[1]**; greater **[1]**; current **[1]**
2. a) potential difference = current × resistance / $V = IR$ **[1]**
   b) 230 = 5 × R **[1]**; $R = \frac{230}{5}$ = 46Ω **[1]**
   c) charge = current × time / $Q = It$ **[1]**
   d) Q = 5 × 120 **[1]**; = 600C **[1]**
3. energy **[1]**; greater **[1]**; current **[1]**
4. a) Closed switch **[1]**
   b) Battery **[1]**
   c) Fuse **[1]**
   d) Light dependent resistor / LDR **[1]**

Page 63 Circuits and Resistance

1. a) To adjust the resistance of the circuit **[1]**; and change the voltage across the component **[1]**
   b) Series **[1]**
   c) Parallel **[1]**
2. Four correctly drawn lines **[3]** (2 marks for two correct lines; 1 mark for one correct line)
   Light dependent resistor (LDR) – Resistance decreases as light intensity increases.
   Thermistor – Resistance decreases as temperature increases.
   Diode – Has a very high resistance in one direction.
   Filament light – Resistance increases as temperature increases.

Answers

Page 64 Circuits and Power

1. a) energy transferred = power × time
   / $E = Pt$ [1]; $E = 1600 × 120$ [1];
   = 192 000J [1]
   b) $\frac{192\,000}{100} × 10$ [1]; = 19 200J [1]
   (accept any equivalent method)
2. a) $V = IR$, $V = 2 × 3$ [1]; = 6V [1]
   b) $18 = 6 + V$ [1]; $V = 18 − 6 = 12V$ [1]

   > The total potential difference in
   > a series circuit is shared across
   > the components.

   c) $18 = 2 × R$ [1]; $R = \frac{18}{2} = 9Ω$ [1]

Page 65 Domestic Uses of Electricity

1. a) 1.5V [1]; d.c. [1]
   b) 230V [1]; a.c. [1]
2. When a device is switched off, the live
   wire before the switch can still be at
   a non-zero potential [1]; touching this
   would create a potential difference
   between the wire and the ground [1];
   this would make current flow through
   the person [1]; which would cause an
   electric shock [1]

Page 66 Electrical Energy in Devices

1. a) Kinetic [1]
   b) It is dissipated / lost [1]; to the
   surroundings [1]
2. a) Energy input = electrical [1]; useful
   energy output = kinetic [1]
   b) It disappears. [1]
   c) 500J [1]
3. The output from the generators goes
   through a step-up transformer [1]; this
   increases the voltage and also reduces
   the current [1]; the low current stops
   the cables from becoming hot [1];
   which means less energy is lost during
   transmission [1]

Page 67 Magnetism and Electromagnetism

1. If free to move, the magnet will rotate
   so that the north pole of the magnet
   [1]; points to the Earth's north pole [1]
2. a) They will repel [1]
   b) It will be attracted to the magnet [1]
3. When the switch is pressed current
   flows in the electromagnet [1]; this
   magnetises the magnet [1]; which
   attracts the armature, causing the
   hammer to hit the gong [1]; the
   movement of the armature breaks the
   circuit, switching off the magnet [1]; the
   armature springs back and remakes the
   circuit, which starts the cycle again [1]

Page 68 Particle Model of Matter

1. a) The energy required [1]; to change
   1kg of a substance from a solid to
   a liquid [1]

   b) thermal energy for a change of
   state = mass × specific latent heat /
   $E = mL$, $E = 0.012 × (2.3 × 10^6)$ [1];
   = 27 600J [1]
2. a) The substance is condensing [1];
   from a gas to a liquid [1]
   b) The particles are slowing down [1];
   and the substance is cooling [1]

Page 69 Atoms and Isotopes

1. a) Electron, −1 [1]; neutron, 0 [1];
   proton, +1 [1]
   b) It has the same number of protons
   [1]; as electrons [1]
   c) ion [1]; positive [1]
2. Path A is a long way from the nucleus
   and the alpha particle goes straight
   through [1]; Path B is close to the
   positive nucleus so the alpha particle is
   deflected [1]; Path C comes very close
   to the nucleus and the alpha particle is
   repelled back the way it came [1]

   > Two positively charged particles
   > will repel each other.

Page 70 Nuclear Radiation

1. a) An unstable atom that gives out
   radiation [1]
   b) Beta decay [1]
   c) It is stable / non-radioactive [1]
   d) Sodium [1]
2. Becquerel / Bq [1]
3. Gamma, beta, alpha [1]

Page 71 Half-Life

1. a) No longer a risk [1]; because it
   has a half-life of just 8 days [1]; so
   would have completely decayed
   to the same level as background
   radiation in the last 30 years [1]
   b) $^{40}_{19}$Potassium [1]; $→ {}^{40}_{20}$Caesium +
   $^{0}_{-1}$e [1] (Accept chemical symbols
   instead of words)
   c) Calculation showing that count
   rate has halved twice in 60 years
   [1]; half-life = 30 years [1]
   d) It has a long range in air [1]; and is
   weakly ionising [1]

Pages 73–86 Biology Practice Exam Paper 1

01.1 Bacterium [1]
01.2 $\frac{70}{14\,000}$ [1]; = 0.005 [1]
02.1 Vomiting [1]; abdominal cramps [1];
02.2 July, August and September [1]
02.3 Incidences higher in the summer
[1]; because food not kept at
a cold enough temperature
in summer / references to
barbeques and undercooked
food [1]
03.1 They can replace the damaged
heart muscle cells. [1]
03.2 Right ventricle [1]

03.3 A build-up of fatty material inside
the coronary arteries [1]; narrows
them down and reduces the flow
of blood [1]; resulting in a lack
of oxygen for the heart muscle [1];
Any three risk factors from:
saturated fat in diet [1]; smoking
[1]; lack of exercise [1]; stress [1];
genetic factors [1]
03.4 Will have the same antigens /
tissue type [1]; so no chance of
rejection [1]
03.5 Stem cells removed from embryos
[1]; the embryos are destroyed /
mention of ethics of disposing of
human embryos [1]
04.1 Fleming [1]; fungus [1]
04.2 4, 3, 1, 2 [3] (1 mark for 4 before 3;
1 mark for 3 before 1; 1 mark for 1
before 2)
05.1 To outcompete other plants [1];
and get more light [1]; so more
photosynthesis / they can produce
more food [1]
05.2 **Any three of:** water enters plant
in the roots [1]; through the root
hairs [1]; passes up the stem [1]; in
the xylem [1]
05.3 The presence of spines [1]; mean
animals are less likely to eat it [1]
05.4 Fewer stomata [1]; so less water
loss [1]; the thick waxy cuticle [1];
means less water is lost through
the epidermis [1]
06.1 Four correctly plotted points [4];
straight line of best fit [1]

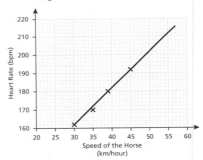

06.2 As speed increases heart rate
increases [1]; straight line (for
these speeds) / use of figures [1]
06.3 A horizontal line should be
drawn on the graph from
200bpm on the y-axis to the line
of best fit and a vertical line
should be drawn down from this
point to the x-axis [1]; accept 48,
49 or 50 km/hour [1]
06.4 glucose [1]; → lactic acid [1]
06.5 Aerobic releases more energy. [1];
Aerobic does not produce lactic
acid. [1]
07.1 C [1]
07.2 Digests food [1]; Kills
microorganisms in food [1]
07.3 Glucose test [1]
07.4 C [1]; D [1]
07.5 Black [1]

07.6 Protease / pepsin / trypsin **[1]**

07.7 Amino acids **[1]**

08.1 water **[1]**; dilute **[1]**; concentrated **[1]**; selectively permeable **[1]**

08.2 Osmosis**[1]**; because the contents of the potato cells were more dilute than the solution **[1]**

08.3 Answer from intercept on graph in the range of 0.35–0.38mol/dm³ **[1]**

08.4 **Top to bottom:** I **[1]**; C **[1]**; D **[1]**; C **[1]**

Pages 87–102 Biology Practice Exam Paper 2

01.1 Fingertip **[1]**

01.2 To make sure the results are reliable **[1]**; because she has a 50/50 chance of guessing right **[1]**

01.3 Less sensitive than fingertip / more sensitive than leg **[1]**; she can tell the difference between one pin and two at 1cm only **[1]**

01.4 She is thinking about it / it is not involuntary **[1]**; it does not involve protecting her body **[1]**

01.5 **Top to bottom:** (A), D, E, B, C **[3]** (2 marks for two correct; 1 mark for one correct)

02.1 **Any two of:** sharp teeth to catch seals **[1]**; white fur for camouflage **[1]**; sharp claws to catch seals **[1]**; forward facing eyes to judge distance **[1]**

02.2 Ursus **[1]**

02.3 Polar bears and Alaskan bears may mate and produce sterile hybrids. **[1]**; The habitats of the polar bears and the Alaskan bears may overlap. **[1]**

02.4 **Any three of:** increased burning **[1]**; of fossil fuels **[1]**; more demand for energy **[1]**; more cars **[1]**; deforestation **[1]**

02.5 **Any three of:** carbon dioxide is a greenhouse gas **[1]**; it traps heat **[1]**; long wavelength energy cannot escape **[1]**; leading to global warming **[1]**

03.1 B **[1]**; B is just downstream from the pile **[1]**

03.2 At B the concentration of copper is higher **[1]**; plants at B are being poisoned by the copper **[1]**

03.3 2 **[1]**

03.4 Plants at B have adapted to living in high copper concentrations **[1]**; plants at A have not experienced high concentrations before **[1]**

03.5 mutation **[1]**; reproduce **[1]**; resistant **[1]**; natural selection **[1]**

04.1 heterozygous **[1]**; tall **[1]**; dominant **[1]**; genotypes **[1]**

04.2 Correctly completed table: with the missing gamete t added **[1]**; correct genotypes in Row 2: Tt, tt **[1]** and correct genotypes in Row 3: Tt, tt **[1]**; 1:1 **[1]**; (Accept a final answer of 50% or $\frac{1}{2}$)

	T	t
t	Tt	tt
t	Tt	tt

05.1 DNA **[1]**; chromosome **[1]**; gene **[1]**

05.2 If Y chromosome is present then it is a boy **[1]**; as boys have XY sex chromosomes **[1]**

05.3 **Any one of:** easier to obtain mother's blood than baby's cells **[1]**; less risk / less likely to cause damage or miscarriage **[1]**

06.1 Asexual **[1]**

06.2 They have exactly the same genes **[1]**

06.3 The bottom part contained the nucleus **[1]**; this means in has the genetic material / DNA / chromosomes **[1]**

07.1 Lining breaks down **[1]**; passes out of the vagina **[1]**

07.2 29th **[1]**

07.3 Ovulation usually occurs around day 14 **[1]**; this would be on the 14th / 15th **[1]**

07.4 Ovulation does not always happen on day 14 / sperm can survive in the female for several days **[1]**

07.5 **Any two of:** pill **[1]**; cap **[1]**; condom **[1]**; coil **[1]**

07.6 Oestrogen **[1]**; progesterone **[1]**

07.7 Ovaries **[1]**

08.1 **Any two of:** body temperature **[1]**; water content **[1]**; blood pH **[1]**; carbon dioxide levels **[1]**; level of urea **[1]**

08.2 A chemical messenger **[1]**; carried in the blood **[1]**

08.3 **Any five of:** need to inject themselves with insulin **[1]**; blood glucose level is too high **[1]**; they cannot produce insulin **[1]**; therefore, they cannot reduce their glucose levels **[1]**; glucose may start to pass out in the urine **[1]**; lead to coma **[1]**

Pages 103–120 Chemistry Practice Exam Paper 1

01.1 Proton +1, Neutron 0, Electron –1 **[1]**

01.2 It has equal numbers of protons and neutrons / equal numbers of positive and negative charges **[1]**

01.3 Mass number 4, Atomic number 2 **[1]**

> The atomic number is the number of protons (and also electrons). The mass number is the total number of particles in the nucleus (protons plus neutrons).

01.4 It has a full outer shell of electrons. **[1]**

01.5 It has the same atomic number. **[1]**; It has a different mass number. **[1]**

02.1 5 **[1]**

02.2 (12 + 16 + 16) = 44 **[1]**

02.3 oxidation–carbon **[1]**; reduction–iron(III) oxide **[1]**

02.4 It is found pure in the ground **[1]**; because it is unreactive **[1]**

03.1 **Any one of:** the density increases as you go down the group **[1]**; the melting point increases as you go down the group **[1]**; the boiling point increases as you go down the group **[1]** (Accept alternative answers, e.g. the density decreases as you go up the group)

03.2 Any answer in the range 200–230°C **[1]**

03.3 cool **[1]**; chlorine **[1]**; liquid **[1]**

03.4 Br_2 **[1]**

03.5 Intermolecular forces break **[1]**

03.6 They all have the same number of electrons (seven) in their outer shell **[1]**

03.7 Sodium ion has no electrons drawn **[1]**; labelled Na⁺ / + ion **[1]**; chlorine ion has eight electrons drawn **[1]**; seven represented by dots and one by a cross **[1]**; labelled Cl⁻ / 1– ion **[1]**

03.8 Boiling point increases **[1]**; Molecular mass increases **[1]**

04.1 Filtration **[1]**; to remove the excess copper(II) oxide **[1]**

04.2 Hazard: **any one of:** chemical on skin **[1]**; chemical in eyes **[1]**; cuts from broken glass **[1]**; Way of reducing the risk: **any one of (as appropriate to hazard):** wash hands **[1]**; wear eye protection **[1]**; ask teacher to clear away broken glass **[1]**

04.3 Metal oxide: calcium oxide **[1]**; acid: hydrochloric acid **[1]**

04.4 $CaCl_2$ **[1]**

05.1 **Similarity: any one of:** they both are made up of carbon atoms **[1]**; both contain strong covalent bonds between carbon atoms **[1]**; **Difference: any one of:** graphite is made up of layers, diamond has no layers **[1]**; graphite contains weak intermolecular forces, diamond only contains strong covalent bonds **[1]**; graphite has delocalised (free) electrons, diamond does not **[1]**

05.2 Electrical conductivity comes from delocalised electrons, which are able to move through the structure **[1]**; this is useful for touch-screens, as they need to be able to conduct electricity to

Answers

work [1]; strength comes from strong covalent bonds between carbon atoms [1]; this is useful for touch-screens so they do not crack / shatter when dropped [1]; graphene is transparent because it is only one atom thick [1]; this is useful for touch-screens so you can see the light coming through from the display underneath [1]

06.1 Battery / cell added and joined to electrodes [1]; electrode connected to positive side of cell labelled anode [1]; electrode connected to negative side of cell labelled cathode [1]

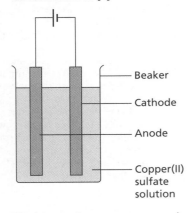

Beaker
Cathode
Anode
Copper(II) sulfate solution

06.2 The ions are free to move around [1]; and conduct electricity [1]

06.3 Copper [1]; it is less reactive than hydrogen [1]

07.1 Hydrochloric acid: 1 [1]; sodium hydroxide: 13 [1]

07.2 Sodium chloride [1]; water [1]

07.3 Add indicator to sodium hydroxide solution [1]; add hydrochloric acid (gradually) [1]; until indicator just changes (colour) / until universal indicator turns green / shows pH7 [1]

07.4 17.5cm³ [1]

07.5 Add the acid in smaller volumes (less than 2.5cm³) [1]; around 15cm³ [1] (accept anything in the range 10–20cm³)

08.1 (s) [1]; (g) [1]

08.2 To let the gas out / to stop the acid spraying [1]

08.3 Sensible scales, using at least half the grid for the points [1]; all points correct [2]; connected by smooth curve [1]

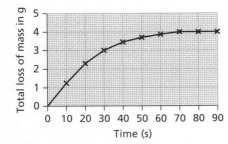

08.4 70s [1]

08.5 All of the acid had reacted [1]

08.6 Hydrogen gas is formed [1]; which escapes into the air [1]

Pages 121–138 Chemistry Practice Exam Paper 2

01.1 2 [1]

01.2 Carbon dioxide – Bubble the gas through limewater [1]; Oxygen – Hold a glowing splint at the open end of a test tube containing the gas [1]

01.3 Carbon particles [1]

01.4 It is colourless [1]; and odourless [1]

01.5 sulfur dioxide – acid rain [1]; carbon particles – global dimming [1]

02.1 Desalination [1]

02.2 It contains other substances [1]; which are dissolved [1] (Accept: it is a mixture for 1 mark)

02.3 Chlorine / ozone / ultraviolet light [1]

02.4 $\frac{250}{1000} \times 1.35$ [1]; = 0.3375 [1]; = 0.34mg [1]

03.1 C_3H_6 [1]

03.2 Add bromine water [1]; turns orange to colourless [1]

03.3 HD poly(ethene) [1]; because it is stronger [1]

03.4 A lot of energy is needed [1]; to break the strong forces between the chains [1]; and allow the chains to move past each other [1]

04 **Effects on the environment: any two of:** melting of ice caps [1]; sea level rise, which may cause flooding and coastal erosion [1]; changes in amount, timing and distribution of rainfall [1]; desertification in some regions [1]; more frequent and severe storms [1]
Effects on people: any two of: flooding of homes [1]; migration of people from affected areas [1]; temperature and water stress [1]; lack of food in some regions [1]
Effects on wildlife: any two of: changes to distribution of species [1]; extinction of some species [1]; temperature and water stress [1]; lack of food in some regions [1]

05.1 They are finite / non-renewable / may run out [1]

05.2 Recycling reduces the amount of metal being mined (metals will last longer) [1]; mining, processing metals and recycling all require energy, which mostly comes from the use of finite resources [1]; collecting and transporting cans uses petrol / diesel [1]; sorting cans and rolling metal blocks requires electrical machinery [1]; melting the cans requires a lot of energy [1]; overall, recycling consumes less energy than producing new cans [1]

06.1 Push the plunger of the syringe down to remove any air in it [1]; the plunger may be pushed out before the end of the reaction [1]; so volume of gas produced not accurately measured [1]

06.2 All points plotted correctly for two sets of data [2]; two lines of best fit drawn [2]

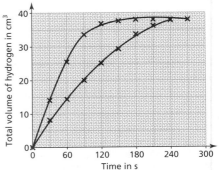

06.3 The higher the concentration, the faster the rate of reaction [1]

06.4 B [1]

06.5 As the concentration increases so does the number of particles of acid in a given volume [1]; so there are more frequent collisions / more collisions per second with magnesium particles [1]; so rate increases / reaction speeds up [1]

06.6 140s [1]

06.7 $\frac{0.88}{140}$ [1]; = 0.0063 [2] (1 mark for 0.006285...)

07.1 2.5 billion years [1]

07.2 40cm³ [1]

07.3 algae [1]; water [1]; photosynthesis [1]

07.4 Carbon dioxide allows short wavelength radiation to pass through [1]; the atmosphere to the Earth's surface [1]; carbon dioxide absorbs outgoing long wavelength radiation [1]

07.5 Increased [1]

07.6 Increased burning of fossil fuels [1]; in vehicle engines / power stations [1] OR increased deforestation [1]; so fewer trees to absorb carbon dioxide from the air [1]

08.1 3 [1]

08.2 15cm (distance moved by S) and 22cm (distance moved by solvent) [1]; $\frac{15}{22}$ = 0.68182 [1]; = 0.68 [1]

08.3 Mobile phase / solvent moves through the paper [1]; and carries different compounds different distances [1]; depending on their attraction for the paper and the solvent [1]

Answers

Pages 139–152 Physics Practice Exam Paper 1

01.1 **From left to right:** 65J [1]; 5J [1]; 30J [1]

01.2 Heat the schools [1]; because it saves the most energy [1]; half of 5J is 2.5J [1]; but a quarter of 65J is over 15J [1]

01.3 efficiency =

$$\frac{\text{useful output energy transfer}}{\text{total input energy transfer}}$$

$\times 100\%, = \frac{30}{100} \times 100\% = 30\%$ [1]

02.1 It spreads out / is dissipated [1]; into the surroundings [1]

02.2 20 + 13 + 7 = 40% [1]

02.3 60% [1]

03.1 work done = force × distance / $W = Fs$, $W = 6 \times 2$ [1]; = 12J [1]

03.2 weight = mass × gravitational field strength / $W = mg$, $m = \frac{6}{10} =$ 0.6kg [1]; gravitational potential energy = mass × gravitational field strength × height / $E_p = mgh$, $E_p = 0.6 \times 10 \times 2 = 12$J [1]

03.3 gravitational [1]; kinetic [1]; sound [1]

04.1 **Top to bottom:** gravitational [1]; light [1]; kinetic [1]; chemical [1]

04.2 Coal and gas [1]

04.3 Nuclear / oil [1]

04.4 Wave / tidal / geothermal / biomass [1]

04.5 Nuclear power stations have high output [1]; and are reliable [1]; **plus any two of:** however they produce nuclear waste [1]; are expensive to build and decommission [1]; have a risk of explosion if something went wrong [1]; are not renewable [1]

05.1 Y = variable resistor [1]; Z = voltmeter [1]

05.2 To control the voltage applied to component X / adjust the resistance [1]

05.3 Voltage [1]

05.4 Current [1]

05.5 Accurately plotted voltage [1]; and current [1]; with all points joined by straight lines [1]

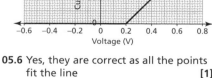

05.6 Yes, they are correct as all the points fit the line [1]

An anomalous result would be significantly higher or lower than the other results or would not fit the pattern.

05.7 A diode [1]

06.1 They have the same number of protons [1]

06.2 Beta decay [1]

06.3 The mass number has not changed during emission, but the proton number has increased by one [1]

06.4 The time it takes for half of the radioactive isotopes to decay / for the count rate to halve [1]

06.5 Because lots (10) half-lives have passed / it has decayed so much [1]; it will no longer be very active / give out much radiation [1]

07.1 kinetic energy = 0.5 × mass × (speed)² / $E_k = \frac{1}{2}mv^2$ [1]; = 0.5 × 1050 × (30 × 30) [1]; = 472 500 [1]; J [1]

07.2 The bus has more mass [1]; so it has more kinetic energy [1]

07.3 0.5 × 1200 × (10 × 10) [1]; = 60 000 [1]; 60 000 − 38 400 = 21 600J [1]

07.4 Because the kinetic energy depends on the square of the speed [1]

08.1 gravitational potential energy = mass × gravitational field strength × height / $E_p = mgh$, $E_p = 65 \times 10 \times 1.25$ [1]; = 812.5J [1]

08.2 His mass is more than 65kg. [1]

08.3 Energy is lost due to air resistance [1]; some kinetic energy is needed to keep moving forward [1]

09.1 12V [1]

09.2 2 + 2 = 4A [1]

09.3 potential difference = current × resistance / $V = IR$, $R = \frac{12}{2}$ [1]; = 6Ω [1]

09.4 It is less / half [1]

09.5 power = potential difference × current / $P = VI$, $P = 12 \times 2$ [1]; = 24W [1]

09.6 $\frac{96\ 000}{(24 \times 2)}$ [1]; = 2000s [1]

Pages 153–166 Physics Practice Exam Paper 2

01.1 So it will be attracted by the magnetic coil [1]

01.2 It will need to increase [1]

01.3 A bigger mass makes a bigger force pulling down on the left, so a bigger force is needed on the right [1]; a bigger current will increase the strength of the magnetic field created by the coil [1]

01.4 The magnet will be stronger [1]; so can balance a bigger mass [1]

02.1 1.2 × 5 = 6 miles [1]

02.2 Displacement is a vector quantity [1]; the cars end up back where they started, so the displacement is zero [1]

02.3 No, the car is changing direction, which requires a force (or it would stay in a straight line) [1] (Accept: changing direction means it accelerates, so has a unbalanced force)

02.4 speed [1]; velocity [1]; direction [1]

02.5 Gravity / the gravitational attraction of the Earth on the Moon [1]

03.1 light = 5×10^{-7}m, UV = 3×10^{-8}m, X-rays = 1×10^{-11}m [2] (1 mark for one in the correct position)

03.2 They all travel at the same speed in a vacuum. [1]; They are all transverse waves. [1]

03.3 **Any pair from:** infrared [1]; used for cooking / remote control [1] **OR** Microwaves [1]; used for communication / cooking [1] **OR** gamma rays [1]; used for sterilising equipment [1]

03.4 wave speed = frequency × wavelength / $v = f\lambda$ [1]; $f = \frac{3 \times 10^8}{300}$ [1]; = 1×10^6 [1]; Hz [1]

03.5 acceleration = $\frac{\text{change in velocity}}{\text{time}}$ / $a = \frac{\Delta v}{t}$ [1]; $a = \frac{10}{4}$ [1]; = 2.5m/s² [1]

03.6 10m/s [1]

03.7 distance travelled = speed × time / $s = vt$ [1]; $s = 10 \times 4 = 40$m [1]

04.1 speed = $\frac{\text{distance travelled}}{\text{time}}$ / $v = \frac{d}{t}$ [1]; $= \frac{(350 - 100)}{(7 - 3)}$ [1]; = 6.5m/s [1]; the car travels at constant speed for the first 3 seconds [1]; it then speeds up to 6.5m/s for 4 seconds [1]; before slowing again for the next 3 seconds [1]

To gain full marks in this type of question, your response must be written using correct grammar and punctuation. Your ideas must be in a sensible order and use the appropriate scientific words.

04.2 acceleration = $\frac{\text{change in velocity}}{\text{time taken}}$ / $a = \frac{\Delta v}{t}$ [1]; = $\frac{30}{6}$ [1]; = 5m/s² [1]

05.1 resultant force = mass × acceleration / $F = ma$ [1]; F = 3500 − 2000 = 1500 [1]; $a = \frac{1500}{1200}$ [1]; = 1.25m/s² [1]

05.2 As the speed increases, the resistive forces increase [1]; until they balance the engine thrust / it reaches terminal velocity [1]

05.3 The distance a car travels between the driver seeing the hazard and applying the brakes [1]

05.4 **Any two of:** fatigue [1]; age [1]; alcohol / drugs [1]; distractions [1] (Accept named distractions, e.g. using a mobile phone)

05.5 Road conditions (Accept a specific adverse condition, e.g. ice, snow, rain, etc.) [1]

Answers

05.6 As the brakes are applied friction occurs between the brakes and the wheels **[1]**; this takes kinetic energy from the car, which is converted to heat **[1]**

06.1 Wavelength **[1]**

06.2 Amplitude **[1]**

06.3 sound **[1]**

06.4 wave speed = frequency × wavelength / $v = f\lambda$ **[1]**; $v = 8 \times 0.015$ **[1]**; = 0.12m/s **[1]**

> 8 waves per second means 8Hz.

06.5 Set up the tank and switch on the stroboscope and adjust the scopes flashing speed so that the waves (or shadows of the waves) appear stationary **[1]**; measure the wavelength with a ruler **[1]**; if using shadows, scale the value based on the total size of shadow compared to the total size of tank **[1]**; flashing lights are a hazard as they can trigger photosensitive epilepsy **[1]**

07.1 They are reversed **[1]**

07.2 They would be closer together / more of them in the same space **[1]**

07.3 A correctly drawn diagram, showing a cylindrical solenoid [1]; with field lines on the outside, looping from one end to the other on both sides (like those around a bar magnet) [1]; and parallel field lines inside, running through the full length of the solenoid [1]

(Direction arrows for current or field are not required)

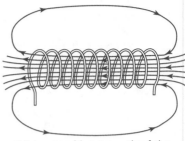

07.4 It increases (the strength of the magnetic field) **[1]**

07.5 C, A, D, B **[3]** (2 marks for starting with C and one other letter in the correct place; 1 mark for starting with C but nothing else correct)

Notes

Notes

Acknowledgements

The author and publisher are grateful to the copyright holders for permission to use quoted materials and images.

Cover & p1 watchara/Shutterstock.com; cover & p1 everything possible/Shutterstock.com; cover Ahuli Labutin/Shutterstock.com; p79 Perfect Lazybones/Shutterstock.com; p80 scenery2/Shutterstock.com; p81 Nadezda Murmakova/Shutterstock.com; p81 Pavel L Photo and Video/Shutterstock.com; p90 Marques/Shutterstock.com; p90 Jim David/Shutterstock.com; p95 National Library Of Medicine/Science Photo Library

Every effort has been made to trace copyright holders and obtain their permission for the use of copyright material. The author and publisher will gladly receive information enabling them to rectify any error or omission in subsequent editions. All facts are correct at time of going to press.

Published by Collins
An imprint of HarperCollinsPublishers Ltd
1 London Bridge Street
London SE1 9GF

© HarperCollinsPublishers Limited 2016

ISBN 9780008326685

Content first published 2016
This edition published 2018

10 9 8 7 6 5 4 3 2 1

British Library Cataloguing in Publication Data.

A CIP record of this book is available from the British Library.

Commissioning Editor: Emily Linnett and Fiona Burns
Biology author: Ian Honeysett
Chemistry author: Gemma Young
Physics author: Nathan Goodman
Project Manager: Rebecca Skinner
Project Editor: Hannah Dove
Designers: Sarah Duxbury and Paul Oates
Copy-editor: Rebecca Skinner
Proofreader: Aidan Gill
Typesetting and artwork: Jouve India Private Limited
Production: Lyndsey Rogers
Printed by Martins the Printers